Springer Theses

Recognizing Outstanding Ph.D. Research

For further volumes:
http://www.springer.com/series/8790

Aims and Scope

The series "Springer Theses" brings together a selection of the very best Ph.D. theses from around the world and across the physical sciences. Nominated and endorsed by two recognized specialists, each published volume has been selected for its scientific excellence and the high impact of its contents for the pertinent field of research. For greater accessibility to non-specialists, the published versions include an extended introduction, as well as a foreword by the student's supervisor explaining the special relevance of the work for the field. As a whole, the series will provide a valuable resource both for newcomers to the research fields described, and for other scientists seeking detailed background information on special questions. Finally, it provides an accredited documentation of the valuable contributions made by today's younger generation of scientists.

Theses are accepted into the series by invited nomination only and must fulfill all of the following criteria

- They must be written in good English.
- The topic should fall within the confines of Chemistry, Physics and related interdisciplinary fields such as Materials, Nanoscience, Chemical Engineering, Complex Systems and Biophysics.
- The work reported in the thesis must represent a significant scientific advance.
- If the thesis includes previously published material, permission to reproduce this must be gained from the respective copyright holder.
- They must have been examined and passed during the 12 months prior to nomination.
- Each thesis should include a foreword by the supervisor outlining the significance of its content.
- The theses should have a clearly defined structure including an introduction accessible to scientists not expert in that particular field.

Wolfgang G. Scheibenzuber

GaN-Based Laser Diodes

Towards Longer Wavelengths and Short Pulses

Doctoral Thesis accepted by
University of Freiburg, Germany

 Springer

Author
Dr. Wolfgang G. Scheibenzuber
Fraunhofer Institute for Applied Solid
  State Physics (IAF)
Tullastraße 72
79108 Freiburg, Germany
e-mail: Wolfgang_Scheib@gmx.de

Supervisor
Prof. Dr. Ulrich T. Schwarz
Department of Microsystems Engineering
University of Freiburg
Georges-Köhler-Allee 106
79110 Freiburg, Germany
e-mail: Ulrich.Schwarz@iaf.fraunhofer.de

ISSN 2190-5053
ISBN 978-3-642-43548-5
DOI 10.1007/978-3-642-24538-1
Springer Heidelberg Dordrecht London New York

e-ISSN 2190-5061
ISBN 978-3-642-24538-1 (eBook)

Printed on acid-free paper

Springer is part of Springer Science+Business Media (www.springer.com)

*A splendid light has dawned on me
about the absorption and emission of
radiation—it will be of interest to you*

Albert Einstein
letter to Michele Besso
November 1916

Supervisor's Foreword

In 2009 the world celebrated 50 years of lasers, bringing to the attention of the public how deeply lasers and laser related technologies have revolutionized both science and everyday life. Next year will be the fiftieth anniversary of the semiconductor laser. Used as a compact light source with high modulation rates, it transformed telecommunications in combination with glass fibers. Today laser diodes are omnipresent in data storage and communication. Still, most of these applications are based on infrared or red laser diodes. The only major application using a short-wavelength laser diode is the Blu-ray Disc, which was enabled by the development of laser diodes based on the gallium nitride material system beginning in 1997. This new material system opened access to the short-wavelength side of the visible spectrum. It also poses new challenges. From the materials physics side it was initially the problem of p-doping and the lack of low-dislocation substrates that posed major obstacles to the development of (Al,In)GaN laser diodes. In terms of semiconductor physics the nitrides puzzled the community with high internal piezoelectric fields and spatial indium fluctuations of the InGaN quantum wells (QWs).

During the last two years great progress has been made towards green-light-emitting laser diodes and short-wavelength, ultrafast laser diodes. In both cases major new applications in the consumer electronics market are the driving force. One is the so-called pico-projector, a device that is small and efficient enough to be part of a handheld, battery-powered device such as a cell phone. These projectors use red, green, and blue laser diodes and will allow us to share images and presentations wherever a white surface is available for projection. It is expected that pico-projectors will become integral to cell phones within a few years, just as cameras are today. The green (Al,In)GaN laser diode is the enabling device for the pico-projector. The other application is again a mass storage device. Sony introduced a new concept for an optical disc based on a picosecond semiconductor laser writing tiny hollow bubbles in the bulk of the material. The prototype used a large-frame, frequency tripled Ti:sapphire laser to write the data onto the disk. The challenge is to develop picosecond (Al,In)GaN laser diodes that can be modulated by an arbitrary bit pattern and with high enough peak power to create the hollow bubbles in the medium. Beyond the consumer market there are many more applications, most prominently in spectroscopy, materials processing, biophotonics, and the life sciences. One example is the field of opto-genetics, where blue, green and red laser diodes are used to stimulate or inhibit—depending on the excitation wavelength—the response of nerve cells. Only semiconductor laser diodes with their tiny footprint will allow this functionality to be integrated with neuro-probes as an interface to the brain.

The research reported in this thesis combines electro-optical characterization with simulations, so as to generate an understanding of the physical mechanisms determining the static and dynamic properties of these (Al,In)GaN laser diodes. This work originated in the environment of cooperations with many academic and industrial partners, through projects funded by the German Research Foundation (DFG), the German government (BMBF), and the European Union. The goals of these projects are a green laser for laser projection on semipolar and c-plane GaN, and the generation of short pulses in the violet to blue spectral region. These joint projects provided access to laser diodes of high quality, which in turn were the enabling factor for the presented measurements of quantities such as optical gain, antiguiding factor, carrier recombination coefficients, and thermal properties with high accuracy.

What can the reader expect from the present thesis? One major contribution to the development of longer-wavelength (Al,In)GaN laser diodes is the characterization of optical gain spectra throughout the spectral range from green to violet and the correct interpretation of optical gain in semipolar laser diodes. A gain model was developed for c-plane, semipolar, and nonpolar InGaN QWs of arbitrary orientation that allows the optical gain spectra to be estimated. The model is based on the $k \cdot p$ approximation of the band structure. The role of anisotropic strain on the piezoelectric field, a self-consistent solution of Schrödinger's and Poisson's equations, and many-body corrections are taken into account to calculate the bound states of the tilted potential of the InGaN QW. The important role of shear strain was pointed out, which causes a switching of the optical polarization of the transition from the conduction band to the topmost valence bands in semipolar QWs at angles close to 45°. Also, it was recognized for the first time that birefringence has a major influence on semipolar (Al,In)GaN laser diodes. These are fundamental results that will remain valid independently of the quality of InGaN QWs, which, in particular for green light emitters, may be improved in the future by different growth techniques.

Since the model has been published, several articles by other groups have appeared that present experimental data supporting the conclusions of the model regarding the dependence of optical gain in semipolar QWs on crystal orientation and polarization switching.

On the way to short-pulse, short-wavelength laser diodes, the first assumption was that (Al,In)GaN laser diodes operate in a similar way to GaAs- or InP-based devices, with some minor corrections due to the shorter wavelength. Indeed, the generation of short pulses by gain switching, active or passive mode-locking, and self-pulsation works in violet laser diodes as well as in the red to infrared spectral region. Also in a configuration with external cavity for linewidth narrowing and tuning, (Al,In)GaN laser diodes behave like the others. So the question is whether the physics of these devices is simply like that of any other separate confinement laser diodes. To a large extent the answer is yes. Yet, it is again the piezoelectric field associated with the InGaN QWs that is responsible for a different physics in ultrafast operation of (Al,In)GaN laser diodes. This field shifts, via the quantum-confined Stark effect (QCSE), not only the gain spectra but also the absorption to higher energies when the field is screened by charger carriers or compensated by the built-in potential of the laser diode's p–n junction. The present thesis shows how to measure the absorption in the absorber of a multi-section laser diode as a function of the applied bias voltage, and how this affects short-pulse operation. Moreover, the dynamical behavior and, in particular, carrier lifetime in the absorber are characterized for the regime of self-pulsation. In this mode of operation, pulses as short as 18 ps with a peak power close to 1 W were achieved. However, the focus is not on record data of short-pulse operation, but on a thorough understanding of the physics necessary to describe and optimize short pulse operation for (Al,In)GaN laser diodes.

In at least one aspect this thesis reaches far beyond laser diodes. It is demonstrated that, when the radiative recombination coefficients are determined from a combination of different characterization methods, carrier injection efficiency and QW inherent loss processes can be separated. While these studies are primarily aimed at an understanding of the dynamical properties of (Al,In)GaN laser diodes, they also allow the Auger coefficient to be measured with 20% accuracy. This result is important for the discussion of the origin of decreasing internal quantum efficiency—also called "efficiency droop"—in light-emitting diodes (LEDs) at high current densities, and therefore a major result for the optimization of high-power LEDs used for solid-state lighting. These LEDs are currently beginning to replace inefficient incandescent and mercury-containing fluorescent lamps. The efficiency droop affects both laser diodes and LEDs at high carrier densities. However, the laser diode is needed in order to distinguish the different mechanisms, because it makes it possible to have high and low photon densities in one device at identical driving conditions, above and below threshold, or in time, before and after the onset of lasing.

I expect that lasers based on the (Al,In)GaN material system will develop from the single in-plane Fabry-Pérot emitter with only moderate output power into a whole family of diode and disc lasers, which will then serve a wide spectrum of applications. Currently the potential of this material system to serve as coherent light sources in the green to ultraviolet spectral region is barely used, compared with the wide variety of red and infrared semiconductor lasers. There have been some demonstrations of distributed feedback (DFB), photonic crystal, and vertical cavity surface emitting (VCSEL) laser diodes. Commercially available Fabry-Pérot laser diodes have also been integrated in external cavity configurations. To generate high optical output power, concepts such as broad area laser diodes and laser arrays were developed. However, in most cases these are just design studies, demonstrating the possibility of a particular concept for (Al,In)GaN but far from a commercial product.

It will take many more years for the (Al,In)GaN material system to reach the maturity of other III-V compound materials with respect to laser physics. This is a question of materials science as well as of the development of high-quality processing techniques and of understanding the complex physics of group-III-nitrides. Devices of semipolar orientation will play a major and growing role alongside c-plane emitters, in particular for longer-wavelength optoelectronics devices. Also dynamic properties and short-pulse operation will continue to be an important topic. The present thesis might serve as reference not only regarding these two aspects of (Al,In)GaN laser diodes.

Freiburg, September 2011   Prof. Dr. Ulrich T. Schwarz

Acknowledgments

This doctoral thesis could never have been successful without the great support of project partners, colleagues, and other researchers. Therefore, I would like to express my sincere gratitude to all who contributed to this work.

First of all, I thank my advisor, Prof. Dr. Ulrich T. Schwarz for his absolute support and experienced advice. The freedom he granted me for my work led to the number of results presented here. Our intense discussions gave me a deeper understanding of the material system (Al,In)GaN and the physics of laser diodes and light emitting diodes. I thank the head of the department "Optoelectronic Modules" at the *Fraunhofer Institute for Applied Solid State Physics (IAF)*, Prof. Dr. Hans-Joachim Wagner for stimulating discussions about diode lasers on GaN and other material systems. I would also like to thank Dr. Klaus Köhler and Dr. Wilfried Pletschen for extensive discussions about epitaxy and processing of group-III-nitride laser diodes and good cooperation within the *Picosecond Challenge* project. The large practical experience they shared with me provided an important link between device physics and technical realizability of laser diodes.

Great thanks go to Prof. Dr. Nicolas Grandjean from *Ecole Polytechnique Fédérale de Lausanne (EPFL)*, and Luca Sulmoni, Dr. Antonino Castiglia, Dr. Jean-Francois Carlin, Dr. Julien Dorsaz from his group, for excellent cooperation within the *Femtoblue* project. Many experiments presented here could not have been done without the high quality laser diode samples grown and processed by them. I also thank Dr. Tim Wernicke and Jens Rass from *TU Berlin* for good cooperation within the *PolarCon* group and for sharing their experimental knowledge on semipolar and nonpolar GaN. I am grateful to Prof. Dr. Bernd Witzigmann from *Kassel University* for stimulating discussions on the calculation of optical gain in GaN-based laser diodes. Further thanks go to Teresa Lermer, Dr. Stephan Lutgen, and Dr. Uwe Strauss from *Osram Opto Semiconductors* for cooperation within the *MOLAS* project and for supplying state-of-the-art violet, blue and green laser diodes. My special thanks go to Prof. Dr. Christian van de Walle for inviting me to give a talk on my research in the nitride seminar of the *University of California Santa Barbara (UCSB)*.

I am thankful to my colleagues of the "Schwarz-Arbeiter" team: Julia Danhof, Hans-Michael Solowan, Lukas Schade, Rüdiger Moser and Christian Gossler, for an excellent working atmosphere and a lot of physical and philosophical discussions. And of course I have to thank my diploma student Christian Hornuss and my bachelor student Helge Höck for taking over a lot of laboratory work.

Finally, I thank my parents, Brigitte and Adolf Scheibenzuber, for their unconditional support and encouragement in all situations. Thank you very much.

Freiburg, September 2011   Wolfgang G. Scheibenzuber

This work has received funding from the European Community's Seventh Framework Programme (FP7/2007-2013, Future and Emerging Technologies-FET) under grant agreement number 238556 (*FEMTOBLUE*). Additional funding has been obtained from the "Fraunhofer Gesellschaft zur Förderung der Angewandten Wissenschaften e. V." within the project *Picosecond Challenge*, the German Federal Ministry for Education and Research (BMBF) within the project *MOLAS* (contract no. 13N9373), and the "Deutsche Forschungsgemeinschaft" (DFG) within the research group *PolarCon* (957).

Contents

Chapter 1
Introduction

Diode lasers are small and efficient sources of laser light. They are well-suited for a wide range of applications, the most popular being optical fiber communications and data storage on CDs and DVDs. The red and infrared laser diodes for these applications are based on the III-IV semiconductor systems AlGaInP and InGaAsP and they are commercially available for a broad spectrum of emission wavelengths. During the last 15 years, remarkable progress was made on short wavelength laser diodes based on the group-III-nitrides. A key advantage of this material system for optoelectronic devices is the wide tuneability of the emission wavelength via the indium content of the InGaN active region. By now, GaN-based laser diodes cover a spectral range from near-UV to green. Since the first demonstration of a room-temperature continuous-wave violet laser diode by Nakamura et al. in 1996 [1] great improvements have been achieved concerning efficiency, device lifetime, output power, and beam quality. Violet laser diodes reach output powers of up to 8W in pulsed operation [2], and the maximum wavelength in continuous wave (cw) operation demonstrated with a GaN-based green laser diode is 525 nm [3].

Besides the application in Blu-ray optical drives, which have a five times higher data density than the DVD, GaN-based laser diodes are suitable for various applications in consumer electronics, optical lithography, sensing, and medical treatment [4]. In particular, the availability of laser diodes for the three basic colors allows the realization of ultra compact, energy efficient laser projectors for integration in mobile devices (see Fig. 1.1). Furthermore, short pulse operations give access to a wider range of additional applications in bio-photonics, like fluorescence lifetime microscopy (FLIM) or fluorescence resonance energy transfer (FRET).

In spite of the achieved progress there remain fundamental issues that limit the accessible range of emission wavelengths and the efficiency of GaN-based laser diodes. Owing to the material properties of the group-III-nitride material system and technological difficulties in the fabrication of epitaxial layers with high indium content, the efficiency of laser diodes and light emitting diodes (LEDs) in the green

W. G. Scheibenzuber, *GaN-Based Laser Diodes*, Springer Theses, DOI: 10.1007/978-3-642-24538-1_1, © Springer-Verlag Berlin Heidelberg 2012

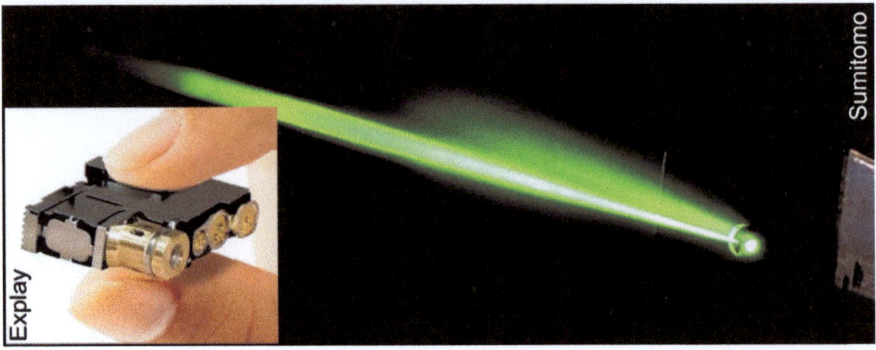

Fig. 1.1 Beam of a green laser diode and photo of a miniature laser projector module (inset)

spectral range is much lower than in the violet to blue spectral range. This phenomenon is generally referred to as the "green gap".

A key issue to overcome the low efficiency of green laser diodes is the optimization of the heterostructure design. Therefore, a detailed understanding of the physical processes that affect the device performance is required. The present work focuses on these physical processes and describes experimental methods to measure the related device parameters which determine the efficiency, such as thermal resistance, optical gain, injection efficiency and recombination coefficients. Simulation approaches are used to understand the qualitative influence of the heterostructure design on these properties. In particular, trends and issues are identified which relate to the increase of the emission wavelength. Furthermore, the concepts developed for the analysis of continuous-wave laser diodes are generalized and applied to picosecond pulse laser diodes with a segmented p-contact design. Short pulse operation is achieved in these multi-section laser diodes by an integrated saturable absorber, which can be tuned by an applied negative bias voltage.

In the introducing chapter, basic properties of the material system (Al,In)GaN and concepts of ridge laser diode design in the group-III-nitrides are presented, which are extended in detail in the subsequent chapters. Band structure and band profile of laser heterostructures are described as well as the optical gain, the mechanism which provides light amplification by stimulated emission. Additionally, a rate equation model is introduced which allows the description of the dynamical properties of laser diodes.

The second chapter treats the influence of device temperature on the properties of a laser diode, especially on the optical gain. A spectroscopic method is demonstrated which allows a precise determination of the thermal resistance and a monitoring of the internal self-heating in pulsed operation.

In the third chapter, optical gain and refractive index spectra of violet, blue and green laser diodes are compared and the low performance of green laser diodes is explained by a reduction and broadening of the gain, which arise from the strong internal piezoelectric fields that occur in strained epitaxial layers with high indium content. The antiguiding factor, an empirical quantity describing the dynamical behavior of laser diodes, is calculated from the quotient of charge-carrier induced refractive index change and differential gain.

Chapter 4 gives a theoretical description of an approach to overcome the decrease of the efficiency with increasing wavelength by growing the heterostructure on a different crystal plane than the commonly used c-plane. Owing to the hexagonal symmetry of wurzite GaN, the growth on such semipolar crystal planes reduces the internal electric fields, which are one cause for the reduction of the optical gain. On semipolar crystal planes, the strain state of the active region is significantly different than on the c-plane. These changes in strain have major consequences for the band structure. The switching of the dominant optical polarization in the emission of particular semipolar devices with increasing indium content, which relates to the shear strain, is explained. Additionally, the influence of birefringence, which occurs in GaN and its alloys, on the optical eigenmodes of the laser waveguide is described. Optical gain spectra for different crystal and waveguide orientations are calculated using a $k \cdot p$-method, and compared to a c-plane laser diode.

In Chap. 5, the dynamics of charge carriers and cavity photons in a laser diode are analyzed using time-resolved spectroscopy with a temporal resolution in the picosecond range. The turn-on behavior of a laser diode in pulsed operation is investigated as a function of pump current and compared to rate equation simulations to extract several device parameters, namely the differential gain, the charge carrier lifetime at threshold, and the gain saturation parameter. By combining the time-resolved measurements with optical gain spectroscopy, a method is developed which allows the determination of the charge carrier recombination coefficients, particularly the Auger coefficient, and the injection efficiency separately.

The Chap. 6 describes the concept of multi-section laser diodes for short pulse generation. Optical gain spectroscopy and time-resolved spectroscopy are used to investigate the absorption spectrum and the charge carrier lifetime in the integrated saturable absorber and their tuneability via the applied negative bias. The rate equation model is generalized to describe the dynamics of multi-section laser diodes and to determine the influence of the absorber properties on the pulse width. The generation of picosecond pulses is demonstrated and the influence of the absorber bias voltage on the pulse width, peak power, and repetition frequency is analyzed.

References

1. S. Nakamura, M. Senoh, S.-I. Nagahama, N. Iwasa, Room-temperature continuous-wave operation of InGaN multi-quantum-well structure laser diodes. Applied Physics Letters **69**(26), 4056–4058 (1996)
2. S. Brueninghoff, C. Eichler, S. Tautz, A. Lell, M. Sabathil, S. Lutgen, U. Strauss, 8 W single-emitter InGaN laser in pulsed operation. Physica Status Solidi A **206**, 1149 (2009)

3. M. Adachi, Y. Yoshizumi, Y. Enya, T. Kyono, T. Sumitomo, S. Tokuyama, S. Takagi, K. Sumiyoshi, N. Saga, T. Ikegami, M. Ueno, K. Katayama, and T. Nakamura, Low threshold current density InGaN based 520–530 nm green laser diodes on semi-polar $(20\bar{2}1)$ free-standing GaN substrates. Appl. Phys. Express 3(12):121001 (2010)
4. A.A. Bergh, Blue laser diode (LD) and light emitting diode (LED) applications. Physica Status Solidi A **201**(12), 2740–2754 (2004)

Chapter 2
Basic Concepts

The design of laser diodes (LDs) in the group-III-nitride material system follows, in principle, concepts already known from other III–V materials. However, the unique properties of the nitrides introduce certain issues that have to be considered in the fabrication of optoelectronic devices.

This chapter starts with a description of the basic layout of an edge-emitting semiconductor laser and explains the functional layers used in such a device. Based thereon, special characteristics of the nitrides, such as the strong piezoelectric polarization and the large difference in mobility for electrons and holes, are explained with respect to their impact on the design of laser diodes. In particular, issues are covered that arise from the usage of layers with high indium content, which are required to reach an emission wavelength in the green spectral range. The mechanism of optical gain in GaN-based laser diodes is explained together with optical losses and their physical origin. Finally, a simulation model based on laser rate equations is introduced which describes the dynamical properties of laser diodes depending on a set of internal device parameters.

2.1 Double Heterostructure Ridge Laser Diodes

The principal requirements for any laser system are a pump source, an active medium which amplifies light via stimulated emission and a resonator for optical feedback. In a double heterostructure ridge laser diode, a layer structure of different semiconductor materials is used to implement these components. Such devices are grown by means of metal-organic vapor phase epitaxy (MOVPE) or molecular beam epitaxy (MBE). The pump mechanism is given by a current which flows through a p–n junction and creates electrons and holes in the space charge region. These charge carriers are trapped in one or multiple thin layers which have a lower bandgap than the surrounding material, the quantum wells. Population inversion, which is required for stimulated emission to overcome the absorption, can be reached in these quantum wells already at moderate pump current densities.

W. G. Scheibenzuber, *GaN-Based Laser Diodes*, Springer Theses,
DOI: 10.1007/978-3-642-24538-1_2, © Springer-Verlag Berlin Heidelberg 2012

An optical waveguide is formed in the transversal direction by employing materials with different refractive index. Additionally, index-guiding of the optical mode in the lateral direction is achieved by fabricating a few micrometer wide ridge on top of the laser diode using photo lithography and dry etching. This way, the emitted light is confined in a small volume around the active region, which provides a high photon density in the active region and thereby an enhancement of the stimulated emission. The cleaved facets of the laser diode chip then form a Fabry–Perot resonator, as they reflect a part of the emitted light back to the material, owing to the refractive index contrast between semiconductor and air. The reflectivities of the facets can be altered by applying high-reflective or anti-reflective dielectric coatings.

A dielectric passivation on the areas aside the ridge limits the current flow to the small ridge volume, where the optical mode is confined. This causes a high current density in the range of kA/cm^2 already at a moderate current and prevents a parasitic current flow in regions where there is little or no photon intensity from the optical mode.

Ridge laser diodes in the group-III-nitride material system are conventionally grown homoepitaxially by MOVPE on c-plane oriented free-standing GaN substrate. Although it is possible to grow such devices heteroepitaxially on sapphire [1] or SiC [2], these substrates have severe drawbacks for the manufacturing of laser diodes, namely a high defect-density and a considerable lattice mismatch to GaN. Best laser performance can be achieved by growth on defect-reduced GaN templates, which are fabricated by epitaxial lateral overgrowth (ELOG) on free-standing GaN wafers [3]. In group-III-nitrides, n- and p-type conductivity are enabled by doping with silicon and magnesium, respectively. The availability of n-doped substrates allows for a vertical current path in GaN-based laser diodes. While the activation energy of Si is lower than the thermal energy at room temperature, it is as high as 170 meV [4] for Mg atoms, which makes high Mg-dopant concentrations in the range of 10^{19} cm^{-3} necessary to achieve sufficient p-type conductivity. Self-compensation effects occuring at such high doping levels limit the maximum achievable p-conductivity in MOVPE-grown p-GaN layers to 1.2 $(\Omega cm)^{-1}$ [5].

Figure 2.1 shows a schematic drawing of a GaN-based ridge laser diode and its epitaxial structure. The optical waveguide is formed by roughly 200 nm of doped GaN between thicker cladding layers of AlGaN, which has a lower refractive index. Situated in the center of the waveguide is the active region, which comprises one or multiple InGaN quantum wells separated by GaN barriers. On the p-side of the active region, a thin AlGaN layer with a high bandgap is implemented, which acts as an electron blocking layer (EBL) to prevent an overflow of electrons into the p-side. It is separated from the quantum wells by an undoped GaN barrier and an undoped spacer layer. A highly Mg-doped GaN layer on top of the p-cladding enables an ohmic contact to the Ni/Au metal stripe on the ridge and the overlying contact pad. The substrate is thinned down to less than 100 μm to facilitate the cleavage of the single chips and an n-contact metal is deposited to the bottom side.

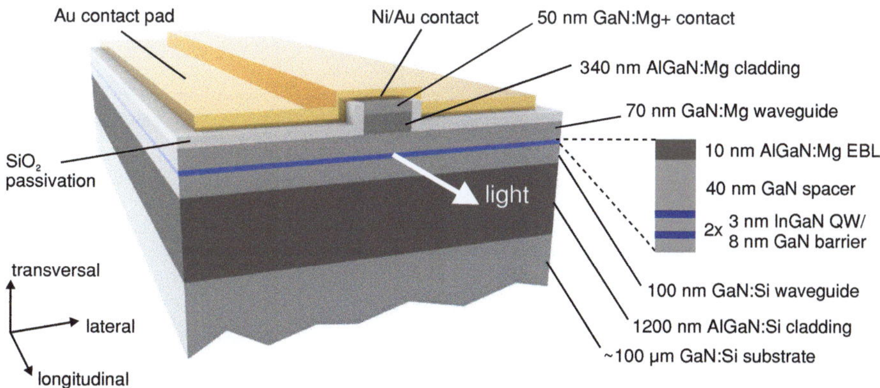

Fig. 2.1 Schematic view of a GaN-based ridge laser diode (not to scale) with description of an example layer structure. The active region is magnified to show the quantum wells

2.2 Heterostructure-Design in Group-III-Nitrides

In order to develop laser diodes which are suitable for applications, tough requirements regarding threshold current, slope efficiency, forward voltage, emission wavelength, output power, and beam quality have to be met. This requires not only a high crystal quality of the epitaxial layers, but also a sophisticated heterostructure design. Particularly in the group-III-nitride system, the width and composition of the individual layers strongly affects the laser performance due to issues that arise from their unique material properties. These issues become even more severe when further developing lasers towards longer emission wavelengths in the green spectral range. The vast number of variation possibilities in a laser diode structure makes it necessary to employ simulation-based concepts for the optimization of heterostructures.

2.2.1 Bandgap and Refractive Index Engineering

The concept of a double heterostructure laser relies on the availability of crystalline materials which differ in bandgap and refractive index, but have a similar lattice constant, so they can be grown pseudomorphically on a common substrate. This allows the growth of optical waveguides and structures that confine the charge carriers, such as quantum wells. Within certain limitations, the ternary alloys AlGaN, AlInN and InGaN fulfill this requirement. Their bandgap and refractive index can be tuned over a wide range by varying their composition, as shown in Fig. 2.2. InGaN is used for the quantum wells. Its bandgap, and thereby the emission wavelength, can be tuned all over the visible spectrum, from near-ultraviolet (GaN) to mid-infrared (InN). Cladding layers for the laser waveguide are typically made of AlGaN, which has a lower refractive index than GaN. Limitations for the epitaxial design of a

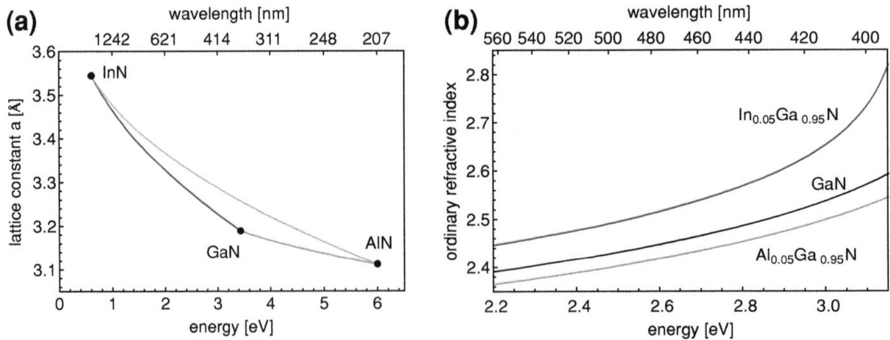

Fig. 2.2 Bandgap and lattice constant a for ternary nitride alloys (**a**) and refractive index as function of photon energy for $Al_{0.05}Ga_{0.95}N$, GaN and $In_{0.05}Ga_{0.95}N$ (**b**). Refractive indices are calculated from an analytical model described in Ref. [6]

laser diode are imposed by the lattice mismatch of the alloys to GaN. The critical thickness of a layer, that is the maximum thickness which can be grown before relaxation effects take place, scales with the inverse of the lattice mismatch. It is thus not possible to stack layers of arbitrary thickness and composition. Exceeding the critical thickness of InGaN leads to the formation of point defects and a rapid deterioration of the optical properties of the layer [7], while for AlGaN it leads to cracking of the epitaxial layers. This is particularly challenging for green laser diodes, as they require both a high indium content in the active region and thick cladding layers, owing to the reduction of the refractive index contrast of GaN and AlGaN with increasing wavelength (compare Fig. 2.2b).

In GaN-based laser diodes, the purpose of the optical waveguide is not only to maximize the overlap of the optical mode with the active region, but also to prevent a leakage of the optical mode to the GaN substrate, which can act as a parasitic second waveguide [8]. In order to optimize the waveguide with respect to these two requirements, numerical simulations can be used that solve the scalar wave equation

$$\left(\frac{\partial^2}{\partial y^2} + \frac{\partial^2}{\partial z^2} + n^2(y,z)k_0^2 \right) E(z) = n_{\text{eff}}^2 k_0^2 E(z), \tag{2.1}$$

which is derived from the Maxwell equations [9]. Here, $n(y,z)$ is the refractive index profile, n_{eff} is the effective index of refraction of the optical eigenmode and $k_0 = 2\pi/\lambda$, with the wavelength in vacuum λ. The transversal and lateral directions are z and y, respectively, and x is the (longitudinal) propagation direction. Figure 2.3 shows one-dimensional optical mode simulations for the layer structure presented in Sect. 2.1, with cladding layers containing 5% aluminum, at different emission wavelengths. While at 405 nm, the optical mode is well confined in the waveguide, with a high intensity at the quantum wells and almost none in the substrate, it becomes significantly broader at 510 nm and coupling to a guided mode in the substrate

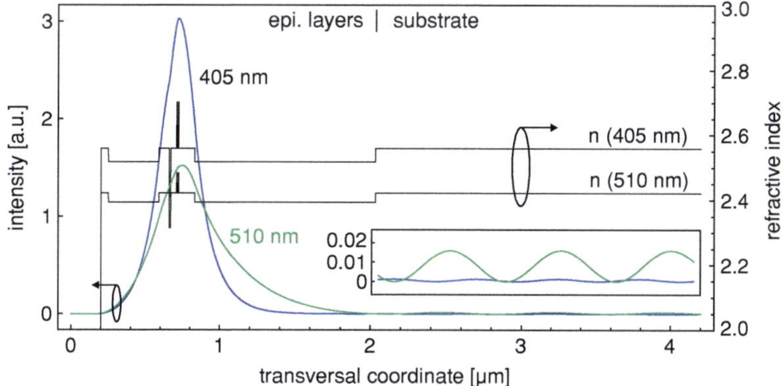

Fig. 2.3 Intensity distribution of optical eigenmodes in the example laser structure shown in Sect. 2.1 at 405 nm (*blue*) and 510 nm (*green*) and corresponding refractive index profiles (*black*). The inset shows a magnification of the intensity distribution in the substrate

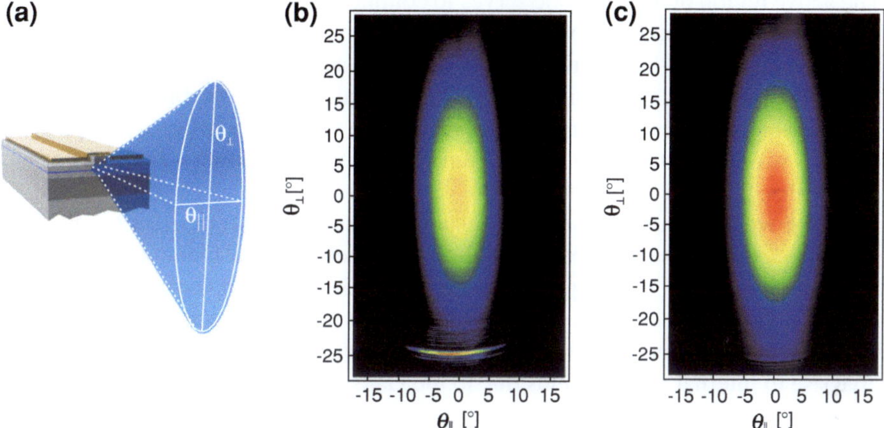

Fig. 2.4 Schematic view of the laser beam (**a**) and far-field of violet laser diodes with (**b**) and without (**c**) a substrate mode peak

occurs. As the far-field of the laser diode is given by the Fourier transform of the optical mode, this substrate mode causes a sharp peak at an angle between 20° and 25° downwards, depending on the effective index of refraction of the laser mode. Insufficient mode confinement also causes an increase in internal optical losses and threshold current [8]. Example far-field patterns of violet laser diodes with different n-claddings, acquired directly with a CCD camera, are shown in Fig. 2.4.

While in violet laser diodes, mode leakage to the substrate can be avoided by using thick claddings with low aluminum content around 5%, this approach becomes increasingly difficult at longer wavelengths. Already in the blue spectral range, the required width of the n-cladding for suffcient substrate mode suppression becomes several micrometers, due to the reduced contrast of refractive index betwen GaN and AlGaN [10]. To avoid substrate modes in green laser diodes and achieve a sufficient beam quality, which is crucial particularly for projection applications, several approaches have been proposed to improve the waveguide: AlInN can be grown lattice matched to GaN and has a much lower refractive index, which makes it well suited as a cladding material [11], although the epitaxial growth of this material with high quality is challenging. Alternatively, InGaN with low indium content can be used for the waveguide layers to improve the refractive index contrast between waveguide and cladding [12]. Another approach employs a highly n-doped GaN layer below the usual AlGaN cladding [13]. Owing to the plasmonic effect, this layer has a reduced refractive index and acts as an additional plasmonic cladding layer.

2.2.2 Piezoelectric Polarization and Active Region Design

The active region of a laser diode serves to confine charge carriers in a small volume where they can recombine via spontaneous or stimulated emission or nonradiative mechanisms. For optimum laser performance, the active region must be designed to maximize the stimulated emission and suppress all other recombination channels. The radiative recombination rate is proportional to the wave function overlap, which depends strongly on the width and indium content of the quantum wells in the group-III-nitrides.

In the wurzite phase, which is the stable phase of the group-III-nitrides, these materials exhibit a strong spontaneous and piezoelectric polarization along the c-axis, which is commonly the growth direction for GaN-based optoelectronic devices. As the polarization depends on the composition and strain state of the material, discontinuities appear at the interfaces of the epitaxial layers. These discontinuities give rise to internal electric fields due to the Gauss law

$$\vec{\nabla} \cdot (\varepsilon \varepsilon_0 \vec{E} + \vec{P}) = 0. \tag{2.2}$$

The internal fields tilt the quantum wells, causing a separation of electron and hole wave functions and a reduction of the transition energy, which is known as the quantum confined Stark effect (QCSE). Although the separated charge carriers can partially screen the internal field, the overlap is still drastically reduced at charge carrier densities relevant for laser operation in quantum wells with an indium content greater than 10%. Increasing the indium content in the quantum wells results in higher strain and stronger internal fields. Wide quantum wells leave electrons and holes more space to separate. Therefore, the wave function overlap reduces with increasing indium conent or QW width, as shown in Fig. 2.5. On the other hand, an increase of the QW width provides a higher overlap of the active region with the optical mode and improves the optical mode confinement due to the high refractive index of InGaN. Still, the strong internal fields limit the range of practical QW widths to few nanometers, at least in conventional devices grown on c-plane GaN.

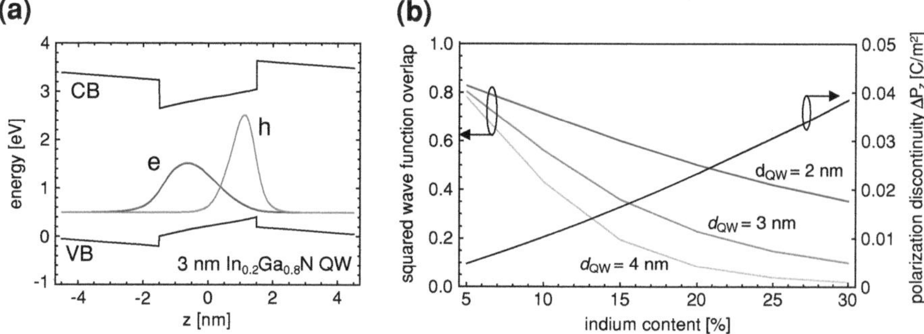

Fig. 2.5 Band profile of a 3 nm In$_{0.2}$Ga$_{0.8}$N quantum well with electron and hole probability densities at a charge carrier density of 5×10^{12} cm^{-2} (**a**). Squared electron-hole wave function overlap for different quantum well widths and polarization discontinuity as functions of indium content (**b**)

Determining the optimum number of quantum wells is difficult for nitride laser diodes. The transparency charge carrier density, which has to be pumped into each quantum well before population inversion starts, is rather high in the nitrides, owing to the high effective hole mass [14]. The low mobility of the holes also leads to an inhomogeneous distribution of charge carriers among the wells, especially for deep quantum wells [15]. This makes a high number of quantum wells undesirable. On the other hand, using only a single quantum well implies a very high charge carrier density in the active region, which causes band filling effects [16] and a low charge carrier lifetime that has to be compensated with a higher pump current. Therefore, the number of QWs is typically chosen between one and three [14, 17, 18].

2.2.3 Band Profile and Charge Carrier Transport

Apart from optimizing the active region for stimulated emission, one also has to ensure that injected charge carriers are captured by the quantum wells and do not overshoot past the active region. In the group-III-nitrides, the mobility of holes is one to two orders of magnitude smaller than the electron mobility [4]. This asymmetry in mobilities causes a significant fraction of electrons to leak out of the active region and reach the p-contact metal. To reduce the charge carrier leakage, an electron blocking layer made of high bandgap material, typically AlGaN, is inserted on the p-side in GaN-based optoelectronic devices. To ensure that this layer blocks electrons and not holes, a high p-doping level is required in the vicinity of the EBL [19]. The injection efficiency is given by the fraction of current which goes into the quantum wells, divided by the total current that flows through the device. For shallow quantum wells, it can be analyzed by calculating the current distribution in the device using a drift-diffusion simulation. The simulation has to take into account the field of the p-n junction, the polarization fields arising from heterostructure interfaces and

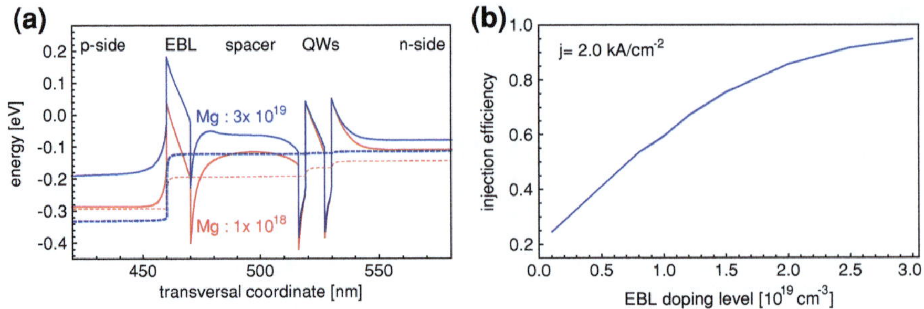

Fig. 2.6 Conduction band profile of the active region and electron blocking layer of a laser diode for Mg doping levels in the EBL of 1×10^{18}cm^{-3} (*red*) and 3×10^{19} cm^{-3} (*blue*) (**a**). The *dashed lines* mark the respective Quasi-Fermi levels. Injection efficiency as a function of EBL doping level at a current density of $2\,\text{kA/cm}^2$ (**b**)

the screening of internal fields by the charge carrier distribution in a self-consistent way. For deeper quantum wells, which are used for green laser diodes, this quasi-equilibrium approach is no longer a good approximation. In such structures, the quantum mechanical scattering rates between bound states in the quantum wells and propagating carriers have to be calculated [15].

Figure 2.6 shows the conduction band profile of a violet laser diode with two quantum wells and an Al$_{0.15}$Ga$_{0.85}$N electron blocking layer, which is calculated using the SiLENSe package [20]. At a low doping level in the EBL, the polarization discontinuity at the heterostructure interface pulls the electron barrier downwards and reduces the effective tunneling barrier for electrons. This reduction cannot be compensated by increasing the aluminum content of the EBL, as a higher aluminum content would also increase the piezoelectric polarization. Instead, the reduction of the barrier can be partly compensated by a high Mg doping level around 2 to 3×10^{19}cm^2 near the EBL. At this doping level, an injection efficiency greater than 90% is achieved at current densities which are typical for laser diodes.

2.3 Band Structure and Optical Gain

The purpose of the active region in a heterostructure laser is to provide optical gain, which amplifies light via stimulated emission. Optical gain is enabled by radiative recombination of electrons and holes from confined states in the quantum wells which have population inversion. To investigate optical gain in a laser diode, it is thus necessary to determine the band structure of the quantum well. Therefore, the 6×6-$k \cdot p$ method is used, a fast numerical method which calculates the energies and wave functions of electrons and holes in vicinity of the Γ-point [21]. This is sufficient for the modeling of optoelectronic devices, as only these states are populated with charge carriers at the relevant injection currents. The energy bands E_i, E_f and wave functions ψ_i, $\vec{\psi}_f$ of initial (i) and final (f) states are obtained from a numerical solution of the Schrödinger equations for valence and conduction band

$$\left(\mathbf{H}_\mathrm{h}(k_x, k_y, -i\frac{\mathrm{d}}{\mathrm{d}z}) + V_{VB}(z)\right) \vec{\psi}_f(z) = E_f \vec{\psi}_f(z), \tag{2.3}$$

$$\left(\mathbf{H}_\mathrm{e}(k_x, k_y, -i\frac{\mathrm{d}}{\mathrm{d}z}) + V_{CB}(z)\right) \psi_i(z) = E_i \psi_i(z). \tag{2.4}$$

Here, $V_{VB/CB}$ are the band profile for valence and conduction band, (k_x, k_y) is the in-plane wavevector of the confined charge carriers and the differential operator d/dz is used to implement quantum confinement along the transversal direction. The operator $\mathbf{H}_h$ is a 6×6 effective mass matrix, which accounts for the $P_{x,y,z}$-angular momentum of the hole states, the anisotropic effective hole masses, the strain potentials and the spin degree of freedom. For the electrons, the effective mass operator $\mathbf{H}_e$ is a scalar, as the electrons are spin degenerate and their angular momentum is S-type. Figure 2.7a shows the band structure of an InGaN quantum well. Three distinct, nearly spin degenerate valence bands occur: The heavy hole (HH), light hole (LH) and crystal field split off (CH) band. Their angular momentum eigenfunctions are $P_x + i P_y$, $P_x - i P_y$ and P_z, respectively. The CH-band is shifted to lower energy due to the relatively large crystal field splitting and a negative strain shift that affects the states which have a hole polarization along to the growth direction. Replicas of these bands appear for the higher order confinement states of the quantum wells. Within the $k \cdot p$-approximation, the band structure of a wurzite semiconductor has a radial symmetry in the c-plane, so the energy dispersion is identical along the x and y-directions in the quantum well.

Radiative recombination has to fulfill energy and momentum conservation, so the relevant transitions in the band structure are vertical, as the momentum of the photon is much smaller than momentum of the charge carriers. Dipole selection rules determine the optical polarization of the emitted photons. As the hole population, which is given by the Fermi-Dirac function, is highest in the topmost bands, this implies that the emitted photons are predominantly polarized in-plane. Therefore, optical gain occurs only for guided modes with an in-plane polarization, the transversal electric (TE) modes.

Within the free-carrier-theory, the material gain G is calculated from the band structure by summing over all possible transitions, weighted with the transition matrix element and a Fermi factor for the population inversion [22],

$$G(\hbar\omega) = \frac{1}{d} \frac{e^2}{4\pi^2 m_0^2 c_0 \varepsilon_0 n_{\mathrm{eff}} \omega E_{\mathrm{hom}}} \sum_{i,f} \int d^2k \left|M_{if}(\vec{k})\right|^2$$
$$\cdot \,\mathrm{sech}\left(\frac{\Delta E_{if}(\vec{k}) - \hbar\omega}{E_{\mathrm{hom}}}\right)\left(f(E_i(\vec{k}), \mu_e) - f(E_f(\vec{k}, \mu_h))\right), \tag{2.5}$$

$$M_{if}(\vec{k}) = \langle i \,|\vec{a} \cdot \vec{p}| \,f \rangle (\vec{k}), \tag{2.6}$$

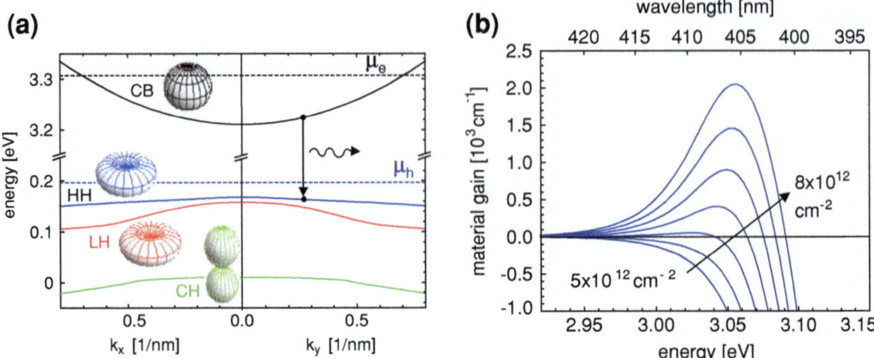

Fig. 2.7 **a** Band structure of a 3 nm $\text{In}_{0.1}\text{Ga}_{0.9}\text{N}$ QW. The insets show the angular momentum eigenfunctions of the individual bands. The *dashed lines* show the quasi-Fermi levels for electrons (*black*) and holes (*blue*) at a charge carrier density of $8 \times 10^{12}\,\text{cm}^{-2}$ and room temperature. The *arrow* marks one possible optical transition. Higher order confinement states appearing between LH and CH band are omitted here for clarity. **b** Homogeneously broadened optical gain spectra calculated from the band structure for charge carrier densities from 5 to $8 \times 10^{12}\,\text{cm}^{-2}$ in steps of $0.5 \times 10^{12}\,\text{cm}^{-2}$ (*bottom to top curve*)

$$\Delta E_{if}(\vec{k}) = E_i(\vec{k}) - E_f(\vec{k}). \tag{2.7}$$

Here, d is the quantum well width, m_0 the free electron mass, c_0 the speed of light in vacuum, ε_0 the vacuum permittivity, e the elementary charge, $\hbar\omega$ the photon energy and n_{eff} the effective index of refraction of the laser mode. Homogeneous broadening due to charge carrier dephasing is implemented by a sech-function, with a broadening energy of typically $E_{\text{hom}} = 25\,\text{meV}$ [23]. For the simulation of real, imperfect quantum wells, an additional inhomogeneous broadening in the range of 30–100 meV is typically employed [24]. M_{if} is the transition matrix element, with the photon polarization vector $\vec{a}$, the momentum operator $\vec{p}$ and the electron and hole wave functions i, f. It is proportional to the wave function overlap in the quantum well. The population inversion of the electron and hole states enters via the difference of the Fermi-Dirac functions $f(E_{i/f}, \mu_{e/h})$, with the quasi-Fermi energies $\mu_{e/h}$. The simple free-carrier gain model presented here is sufficient to study qualitatively the dependence of optical gain on the device design. To reach quantitative agreement with experiments, more sophisticated models have to be used that take into account many-body effects [22].

Calculated gain spectra for different charge carrier densities are shown in Fig. 2.7b. At a sufficiently high charge carrier density, positive gain occurs at photon energies close to the effective bandgap, which is given by the bandgap of bulk InGaN plus the strain shift, the confinement energy of the charge carriers and the redshift due to the quantum confined Stark effect. The spectral width of the optical gain is determined by the homogeneous and inhomogeneous broadening and the filling of energetic states in the band structure. At a photon energy much smaller than the effective bandgap, the gain is zero and there is no absorption by the quantum wells, as no such energetic transitions are available in the band structure. Positive gain is reached only if there are possible transitions in the band structure which have population inversion. The population of a state in a semiconductor is given by the Fermi-Dirac distribution function $f(E, \mu)$, so the condition of population inversion translates into a condition for the transition energy $\hbar\omega$:

$$\underbrace{\left(1 + \exp(\beta(E_f(k_{tr}) - \mu_h))\right)^{-1}}_{=f(E_f(k_{tr}),\mu_h)} < \underbrace{\left(1 + \exp(\beta(E_i(k_{tr}) - \mu_e))\right)^{-1}}_{=f(E_i(k_{tr}),\mu_e)} \tag{2.8}$$

$$\Rightarrow \quad \mu_e - \mu_h > E_i(k_{tr}) - E_f(k_{tr}) = \hbar\omega$$

with $\beta = 1/(k_B T)$, k_B the Boltzmann constant, T the temperature (which is 300 K in this case), $E_{i/f}(k)$ the conduction and valence band dispersion relations and k_{tr} the transition wavenumber. Optical gain is thus only possible for transitions which fulfill $\hbar\omega < \mu_e - \mu_h$. For energies higher than the difference of the quasi-Fermi levels $\mu_e - \mu_h$, there is no population inversion and the quantum well becomes absorbing.

The material gain can be implemented into a waveguide simulation as the imaginary part of the refractive index of the quantum well layers. The modal gain g of the guided mode is then obtained from the imaginary part of the calculated effective index of refraction. As an approximation, it is also possible to estimate the modal gain as the product of the material gain and the optical confinement factor Γ, which is the overlap integral of the optical mode with the quantum wells:

$$g(\hbar\omega) = \Gamma G(\hbar\omega) = \left(\int_{\text{QWs}} |E(z)|^2 dz\right) G(\hbar\omega) \tag{2.9}$$

In a laser diode waveguide, there is not only optical gain but also losses due to absorption on dopant atoms or crystal defects. For nitride based laser diodes, the dominant loss mechanism is absorption on bound holes from non-ionized acceptors in the p-type layers [25]. The fraction of non-ionized acceptors in these layers is high due to the large acceptor activation energy of the Mg atoms. The optical losses of Mg-doped GaN are 50–$100\,\text{cm}^{-1}$ [26]. Therefore it is important to minimize the overlap of the optical mode with p-type layers to achieve low absorption losses [27].

These internal losses are described by an absorption coefficient α_{int}. Another loss mechanism is the transmission of photons through the front and back mirrors. The propagation of a monochromatic electromagnetic wave is described by a complex effective index of refraction n_{eff},

$$E(x) = E_0 e^{i n_{\text{eff}} k_0 x}, \tag{2.10}$$

$$n_{\text{eff}} = n_R + i\frac{\alpha}{2k_0}. \tag{2.11}$$

Here, n_R is the real part of the effective index of refraction and α is the extinction or amplification coefficient (depending on whether it is positive or negative). The intensity after one round trip in the laser resonator is:

$$I(2L) = R_1 R_2 I_0 e^{(g(\hbar\omega)+\alpha_{\mathrm{int}})2L}, \tag{2.12}$$

with the front and back mirror reflectivities R_1 and R_2, the resonator length L, the modal gain $g(\hbar\omega)$ and the internal losses α_{int}. Combining Eqns. 10 to 12 gives the extinction/amplification coefficient α:

$$\alpha = -g(\hbar\omega) + \alpha_{\mathrm{int}} + \underbrace{\frac{1}{2L} \ln(R_1 R_2)}_{\alpha_{\mathrm{m}}}. \tag{2.13}$$

At the laser threshold, the intensity inside the resonator remains constant. The optical gain then just compensates the internal losses α_{int} and the losses of light which is coupled out at the mirrors α_{m}. Therefore, α must become zero, which gives the threshold gain g_{th}:

$$g_{\mathrm{th}} = \alpha_{\mathrm{int}} + \alpha_{\mathrm{m}}. \tag{2.14}$$

To reach a low laser threshold, it is thus necessary to minimize the optical losses in the resonator and to achieve a high optical gain from the quantum wells per injected charge carrier.

2.4 Laser Dynamics

At a sufficiently high pump level, laser operation is enabled by the interaction of the photon field in the cavity and the charge carrier reservoir in the active region of the laser diode. This interaction can be modeled by a system of two coupled differential equations for the charge carrier number N and the photon number S, the rate equations [28]:

$$\frac{\mathrm{d}N}{\mathrm{d}t} = \frac{\eta_{\mathrm{inj}} I}{e} - \frac{N}{\tau(N)} - g(N, S)cS, \tag{2.15}$$

$$\frac{\mathrm{d}S}{\mathrm{d}t} = g(N, S)cS + \beta B N^2 - (\alpha_{\mathrm{int}} + \alpha_{\mathrm{m}})cS. \tag{2.16}$$

Here, I is the pump current, η_{inj} the injection efficiency, $g(N, S)$ the modal gain, $c = c_0/n_{\mathrm{gr}}$ the speed of light in the waveguide with the group refractive index n_{gr}, B is the coefficient for spontaneous emission, $\tau(N)$ is the charge carrier lifetime and β the ratio of spontaneous emission into the laser mode, which is estimated as 1×10^{-5} [29].

The notation presented here works with charge carrier and photon numbers instead of densities to keep geometrical factors out of the calculation. The model can be alternatively formulated in terms of area (2D) or volume (3D) densities by scaling the parameters with the active region area or volume. Within this simplified model,

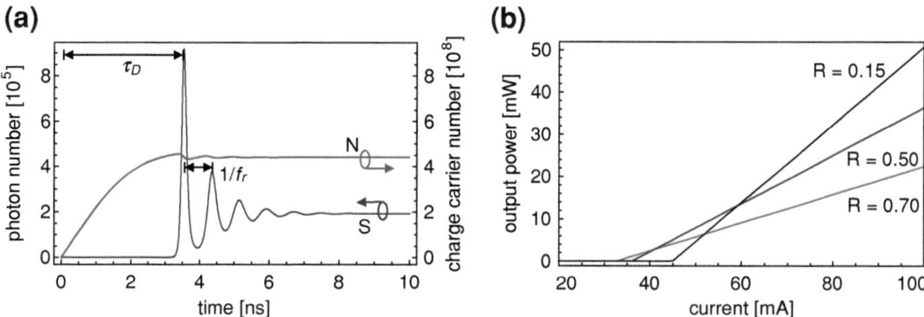

Fig. 2.8 **a** Example time-dependent solution of rate equations for the charge carrier number N and the photon number S for an electrical square pulse starting at $t = 0$. The *arrows* mark the turn-on delay τ_D and the inverse oscillation frequency $1/f_r$. **b** Simulated output power as a function of pump current for different front mirror reflectivites and a cavity length of $L = 600\,\mu$m

the wavelength dependency of the parameters is neglected and single-mode operation is assumed. Spectral dynamics like mode hopping and multi-mode emission can be implemented by replacing Eq. 2.16 by a set of equations for the individual longitudinal modes with appropriate coupling terms [30].

The dependency of the charge carrier lifetime τ on N can be approximated by the ABC-model

$$\frac{N}{\tau(N)} = R(N) = R_{\text{SRH}} + R_{\text{spont}} + R_{\text{Aug}} =$$
$$= AN + BN^2 + CN^3$$

(2.17)

where the first term relates to Shockley-Read-Hall (SRH) recombination on defects, the second term corresponds to the spontaneous emission and the third term implements Auger recombination. The functional dependence of the modal gain $g(N, S)$ on the charge carrier and photon number can be approximated by a linear model [28],

$$g(N, S) = \frac{\mathrm{d}g}{\mathrm{d}N} \frac{(N - N_{\text{tr}})}{1 + k_{\text{sat}}S},$$

(2.18)

where $\mathrm{d}g/\mathrm{d}N$ is the differential gain per carrier number, N_{tr} is the transparency carrier number, and k_{sat} implements gain saturation due to spectral hole burning.

A time-dependent solution of the rate equation model allows to study the dynamical behavior of the laser diode upon turn-on or fast modulation, while the stationary solution yields the dependence of the output characteristics on internal device parameters. Figure 2.8a shows the time evolution of charge carrier number and photon number upon turn-on. After the onset of the electric pulse, there is a turn-on delay τ_D of few nanoseconds during which the active region is filled with carriers up to threshold, then the optical output starts. The laser diode exhibits relaxation oscillations with a frequency f_r and reaches steady state after several nanoseconds. Turn-on delay and relaxation frequency depend on pump current and on device properties such as the carrier lifetime at threshold and the differential gain. In steady state laser operation, the charge carrier number is clamped to its threshold value and does not depend on the pump current.

Solving Eqs. 2.15 and 2.16 for steady state conditions relates the injection efficiency to the slope of the power-vs.-current curve and the cavity losses

$$\frac{dP}{dI} = \eta_{inj}\frac{\alpha_m}{\alpha_m + \alpha_{int}}\frac{\hbar\omega}{e}. \tag{2.19}$$

This provides an accurate method to determine the injection efficiency η_{inj} from measurements of the output power and the internal losses. The influence of the front mirror reflectivity on the output characteristics is shown in Fig. 2.8b. The reflectivity of the rear mirror is set to 1. A high front mirror reflectivity means low mirror losses and therefore a lower threshold current, at the cost of a small slope efficiency dP/dI above threshold. Reducing the mirror reflectivity improves the slope efficiency, as more photons are coupled out of the cavity, but also increases the threshold.

References

1. S. Nakamura, M. Senoh, S.-I. Nagahama, N. Iwasa, Room-temperature continuous-wave operation of InGaN multi-quantum-well structure laser diodes. Appl. Phys. Lett. **69**(26), 4056–4058 (1996)
2. S. Bader, B. Hahn, H. Lugauer, A. Lell, A. Weimar, G. Brüderl, J. Baur, D. Eisert, M. Scheubeck, S. Heppel, A. Hangleiter, V. Härle, First european GaN-based violet laser diode. Physica Status Solidi A **180**, 177–182 (2000)
3. K. Motoki, Development of gallium nitride substrates. SEI Tech. Rev. **70**, 28–35 (2010)
4. J. Piprek, S. Nakamura, Physics of high-power InGaN/GaN lasers. IEE Proc. Optoelectron. **149**(4), 145–151 (2002)
5. U. Kaufmann, P. Schlotter, H. Obloh, K. Köhler, M. Maier, Hole conductivity and compensation in epitaxial GaN:Mg layers. Phys. Rev. B **62**(16), 867–872 (2000)
6. R. Goldhahn, C. Buchheim, P. Schley, A. Winzer, H. Wenzel, Optical Constants of Bulk Nitrides. In: J. Piprek (eds) *Nitride Semiconductor Devices: Principles and Simulations, chapter 5* (Wiley VCH, Weinheim, 2007) pp. 95–116
7. D. Queren, M. Schillgalies, A. Avramescu, G. Brüderl, A. Laubsch, S. Lutgen, U. Strauss, Quality and thermal stability of thin InGaN films. J. Cryst. Growth **311**(10), 2933–2936 (2009)
8. V. Laino, F. Roemer, B. Witzigmann, C. Lauterbach, U.T. Schwarz, C. Rumbolz, M. Schillgalies, M. Furitsch, A. Lell, V. Härle, Substrate modes of (Al, In)GaN semiconductor laser diodes in SiC and GaN substrates. IEEE J. Quantum Electron. **43**, 16–24 (2007)
9. C. Henry, B. Verbeek, Solution of the scalar wave equation for arbitrarily shaped dielectric waveguides by two-dimensional Fourier analysis. J. Lightwave Technol. **7**(2), 308–313 (1989)
10. U. Strauss, C. Eichler, C. Rumbolz, A. Lell, S. Lutgen, S. Tautz, M. Schillgalies, S. Brüninghoff, Beam quality of blue InGaN laser for projection. Physica Status Solidi C **5**(6), 2077–2079 (2008)
11. A. Castiglia, E. Feltin, G. Cosendey, A. Altoukhov, J.-F. Carlin, R. Butte, N. Grandjean, Al0.83In0.17N lattice-matched to GaN used as an optical blocking layer in GaN-based edge emitting lasers. Appl. Phys. Lett. **94**(19), 193506 (2009)

12. T. Lermer, M. Schillgalies, A. Breidenassel, D. Queren, C. Eichler, A. Avramescu, J. Müller, W. Scheibenzuber, U. Schwarz, S. Lutgen, U. Strauss, Waveguide design of green InGaN laser diodes. Physica Status Solidi A **207**(6), 1–4 (2010)
13. P. Perlin, K. Holc, M. Sarzynski, W. Scheibenzuber, L. Marona, R. Czernecki, M. Leszczynski, M. Bockowski, I. Grzegory, S. Porowski, G. Cywinski, P. Firek, J. Szmidt, U. Schwarz, T. Suski, Application of a composite plasmonic substrate for the suppression of an electromagnetic mode leakage in InGaN laser diodes. Appl. Phys. Lett. **95**(26), 261108 (2009)
14. S.-N. Lee, H.Y. Ryu, H.S. Paek, J.K. Son, T. Sakong, T. Jang, Y.J. Sung, K.S. Kim, K.H. Ha, O.H. Nam, Y. Park, Inhomogeneity of InGaN quantum wells in GaN-based blue laser diodes. Physica Status Solidi C **4**(7), 2788–2792 (2007)
15. D. Sizov, R. Bhat, A. Zakharian, K. Song, D. Allen, S. Coleman, C. Zah, Carrier transport in InGaN MQWs of aquamarine- and green-laser diodes. IEEE J. Select. Top. Quantum Electron. (2011, to be published)
16. W.G. Scheibenzuber, U.T. Schwarz, T. Lermer, S. Lutgen, U. Strauss, Thermal resistance, gain and antiguiding factor of GaN-based cyan laser diodes. Physica Status Solidi A **208**, 1600 (2011)
17. D. Queren, A. Avramescu, M. Schillgalies, M. Peter, T. Meyer, G. Brüderl, S. Lutgen, U. Strauss, Epitaxial design of 475 nm InGaN laser diodes with reduced wavelength shift. Physica Status Solidi C **4**, 1–4 (2009)
18. S. Uchida, M. Takeya, S. Ikeda, T. Mizuno, T. Fujimoto, O. Matsumoto, T. Tojyo, M. Ikeda, Recent progress in high-power blue-violet lasers. IEEE J. Select. Top. Quantum Electron. **9**(5), 1252–1259 (2003)
19. S. Grzanka, G. Franssen, G. Targowski, K. Krowicki, T. Suski, R. Czernecki, P. Perlin, M. Leszczynski, Role of the electron blocking layer in the low-temperature collapse of electroluminescence in nitride light-emitting diodes. Appl. Phys. Lett. **90**(10), 103507 (2007)
20. STR Group Ltd., Simulator of Light Emitters based on Nitride Semiconductors (SiLENSe). http://www.semitech.us/products/SiLENSe
21. S. Chuang, C. Chang, $k \cdot p$ method for strained wurtzite semiconductors. Phys. Rev. B **54**(4), 2491–2504 (1996)
22. S.W. Chow, W.W. Koch, *Semiconductor-Laser Fundamentals* (Springer, Berlin, 1998)
23. B. Witzigmann, V. Laino, M. Luisier, U. Schwarz, H. Fischer, G. Feicht, W. Wegscheider, C. Rumbolz, A. Lell, V. Härle, Analysis of temperature-dependent optical gain in GaN-InGaN quantum-well structures. IEEE Photonics Technol. Lett. **18**(15), 1600–1602 (2006)
24. K. Kojima, U.T. Schwarz, M. Funato, Y. Kawakami, S. Nagahama, T. Mukai, Optical gain spectra for near UV to aquamarine (Al, In)GaN laser diodes. Opt. Express **15**, 7730–7736 (2007)
25. E. Kioupakis, P. Rinke, C.G. Van de Walle, Determination of internal loss in nitride lasers from first principles. Appl. Phys. Express **3**(8), 082101 (2010)
26. M. Kuramoto, C. Sasaoka, N. Futagawa, Reduction of internal loss and threshold current in a laser diode with a ridge by selective re-growth (RiS-LD). Physica Status Solidi A **334**(2), 329–334 (2002)
27. U.T. Schwarz, M. Pindl, E. Sturm, M. Furitsch, A. Leber, S. Miller, A. Lell, V. Härle, Influence of ridge geometry on lateral mode stability of (Al, In)GaN laser diodes. Physica Status Solidi A **202**(2), 261–270 (2005)
28. K. Petermann, *Laser Diode Modulation and Noise* (Kluwer Academic Publishers, Dordrecht, 1991)
29. L.A. Coldren, S.W. Corzine, *Diode Lasers and Photonic Integrated Circuits* (Wiley, New York, 1995)
30. B. Schmidtke, H. Braun, U.T. Schwarz, D. Queren, M. Schillgalies, S. Lutgen, U. Strauss, Time resolved measurement of longitudinal mode competition in 405 nm (Al, In)GaN laser diodes. Physica Status Solidi C **6**, S860–S863 (2009)

Chapter 3
Thermal Properties

Many applications of GaN-based laser diodes require a stable output in pulsed or fast modulated operation and a high output power. Self-heating of the devices due to ohmic losses and non-radiative recombination is a limiting factor, as an increase in device temperature results in a reduction of output power, a redshift of the emission wavelength, and thermal lensing. Therefore it is crucial to investigate the mechanisms of self-heating in laser diodes.

In this chapter, the influence of the device temperature on the laser output is discussed, and a precise method to determine the thermal resistance of laser diodes based on high resolution electroluminescence spectroscopy is presented. Furthermore, time resolved spectroscopy is used to investigate the dynamics of self-heating. From the time evolution of the internal temperature, heating contributions in different subsystems can be distinguished and their respective heat capacity can be estimated.

The sample investigated in this chapter is a commercial blue laser diode mounted in a TO-56 can from *Nichia Corporation* [1]. It is grown on free-standing GaN substrate by metal-organic vapor phase epitaxy. At 23°C heat sink temperature, its threshold current is 26 mA at a forward voltage of 4.2 V and the slope efficiency is 0.51 W/A. The emission wavelength is 443 nm.

3.1 Temperature Dependence of Output Characteristics

The internal temperature of a laser diode strongly influences its output power and emission wavelength. An increase in temperature flattens the Fermi–Dirac partitions for electrons and holes in the quantum wells and thereby reduces the degree of population inversion, which leads to a decrease in optical gain at a given charge carrier density. Consequently, a higher pump level is required to compensate the optical losses, and the threshold current increases. Additionally, the fraction of hot electrons which can overcome the electron blocking layer increases, reducing the injection efficiency and thus also the slope efficiency. At high input currents in the range of typically several hundred mA, a thermal roll-over occurs in continuous

W. G. Scheibenzuber, *GaN-Based Laser Diodes*, Springer Theses,
DOI: 10.1007/978-3-642-24538-1_3, © Springer-Verlag Berlin Heidelberg 2012

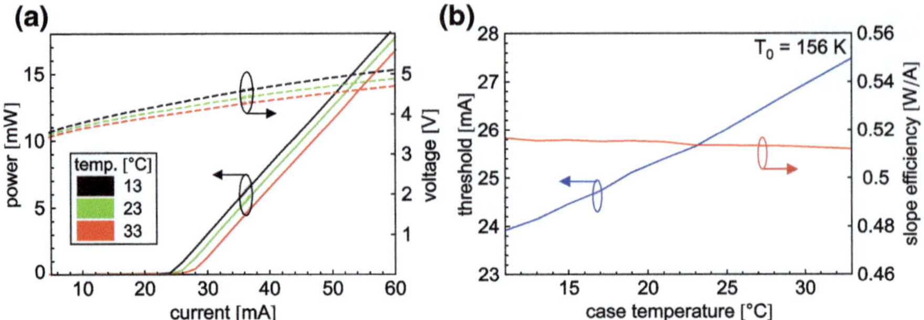

Fig. 3.1 Output power (*solid lines*) and forward voltage (*dashed lines*) in continuous-wave operation as functions of current at different case temperatures (**a**), threshold current (*blue*) and slope efficiency (*red*) as functions of case temperature (**b**)

wave (cw) operation, which limits the maximum output power. The forward voltage slightly decreases with increasing temperature due to a thermal activation of Mg acceptors. This effect partly compensates the reduction of overall power efficiency caused by the deterioration of threshold and slope efficiency.

Figure 3.1 shows output power and forward voltage characteristics for different case temperatures. For this measurement, the laser diode is mounted to a copper heat sink, the temperature of which is controlled by Peltier elements. Assuming that in thermal equilibrium the self-heating causes a constant temperature difference between heat sink and laser wave-guide, the changes in heat sink temperature translate directly to changes in the internal temperature. For this laser diode, the threshold increases slightly from about 24 mA at 11°C to 27 mA at 33°C, while the slope efficiency remains almost constant. The dependency of the threshold current I_{th} on the temperature is typically described by an empirical relation [2]

$$I_{th} = I_0 \exp(T/T_0), \tag{3.1}$$

with a characteristic temperature T_0, which is in this case 156 K.

The temperature of the laser diode also affects its emission spectrum. Figure 3.2a shows spectra of the blue laser diode at a current of 60 mA and heat sink temperatures from 13 to 23°C. The laser diode emits on several longitudinal modes of the laser resonator, which is typical for Fabry-Perot type laser diodes, as the gain spectrum is much broader than the spectral separation of the modes. Two distinct features can be identified in this measurement: The individual longitudinal modes shift by 15 pm/K, while the center wavelength, which is given by the first moment of the spectrum, shifts by 36 pm/K. Similar values for both shifts have been measured in Ref. [3]. The shift of the longitudinal modes relates to the temperature dependence of the modal refractive index of the wave-guide [4]. On the other hand, the center wavelength is determined by the peak of the optical gain spectrum, which shifts to red upon heating due to the temperature dependence of the bandgap [5] and the dependence of the population inversion on the temperature of the charge carrier plasma via the Fermi–Dirac distribution.

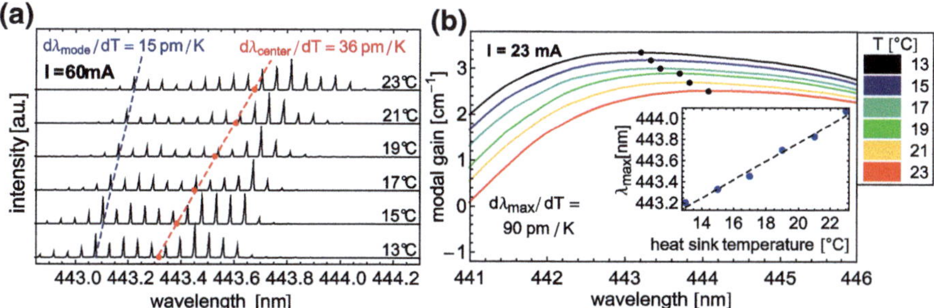

Fig. 3.2 **a** cw spectra above threshold at 60 mA for different heat sink temperatures. *Red points* mark the center wavelength for each spectrum. The *dashed blue* and *red lines* follow the shift of the longitudinal modes and the center wavelength, respectively. **b** Optical gain spectra at 23 mA for different heat sink temperatures. Points indicate the maximum of each curve. The inset shows a linear fit (*dashed line*) of the maximum position *versus* temperature (*points*)

To prove that the shift of the center wavelength corresponds to a shift of the gain, the optical gain spectrum is measured slightly below threshold at different heat sink temperatures using the Hakki-Paoli method [6] (the application of the Hakki-Paoli method to GaN-based lasers is explained in Sect. 4.2). As can be seen in Fig. 3.2b, the peak of the gain spectrum shifts considerably to red with increasing temperature. This redshift is even more pronounced than the shift of the center wavelength in Fig. 3.2a, as the increase in temperature also leads to a reduction of the gain at a given charge carrier density. Above threshold, the charge carrier density has to be increased with increasing temperature to compensate this reduction of gain, shifting the peak slightly to blue again and therefore reducing the observed net redshift.

3.2 Thermal Resistance

Especially for high power laser diodes which operate at high pump currents it is crucial to optimize the heat transport in the device to keep the temperature-related deterioration of the output power as low as possible. The heat transport in a semiconductor chip, as illustrated in Fig. 3.3, is determined by the thermal resistance R_{th}, which is proportional to the temperature offset between junction and heat sink:

$$\Phi_{\text{th}} = \Delta T / R_{\text{th}} = P_{\text{in}} - P_{\text{opt}}, \tag{3.2}$$

where Φ_{th} is the heat flux out of the device, which is equal to the electrical input power P_{in} minus the optical output power P_{opt}.

The thermal resistance can be measured via electroluminescence spectroscopy by making use of the fact that above the laser threshold, the charge carrier density in the active region N is pinned, which means it stays nearly constant while the pump current is increasing. The effective index of refraction of the laser cavity is a function

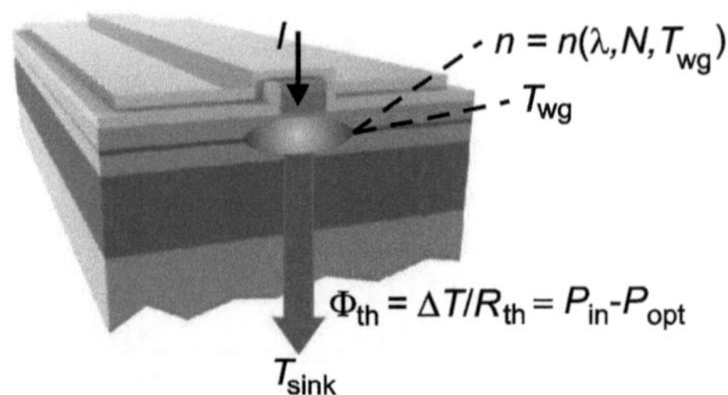

Fig. 3.3 Schematic drawing of the heat transport in a ridge laser diode

of N, the waveguide temperature T_{wg} and the photon wavelength λ:

$$n = n(\lambda, N, T_{wg}). \tag{3.3}$$

When the internal temperature changes, the longitudinal mode spectrum shifts due to the temperature dependence of the effective refractive index of the laser resonator. Above threshold, N is almost constant, so the refractive index change is mainly a function of the waveguide temperature. To determine the thermal resistance, a small part of the longitudinal mode spectrum is recorded at threshold, then the current is slightly increased, which leads to a thermal change of the refractive index and thus a change of the position of the longitudinal modes (see Fig. 3.4a). Then, the heat sink temperature is decreased until the initial mode position is restored. A constant position of the longitudinal modes corresponds to a constant internal temperature. Repeating this procedure for several pump currents and linear fitting the heat sink temperature with constant longitudinal mode position versus the dissipated power $P_{in} - P_{opt}$ gives the thermal resistance R_{th}, as shown in Fig. 3.4b. For the investigated sample, the thermal resistance is 35 ± 1 K/W, where the small measurement error of this method is due to the high spectral resolution of about 4 pm of the experimental setup.

3.3 Dynamics of Self-Heating

In continuous-wave operation, the laser diode is in thermal equilibrium and the offset of internal and case temperature is given by the product of thermal resistance and dissipated power. When operated in pulsed mode, a laser diode does in general not reach thermal equilibrium. The magnitude of self-heating then depends on the pulse width, duty cycle and the thermalization time τ_{th}, which is given by the product of thermal resistance and heat capacity c_{th}

$$\tau_{th} = R_{th} c_{th}. \tag{3.4}$$

As shown in Sect. 3.1, the emission spectrum of a laser diode depends strongly on its internal temperature. Therefore, time resolved spectroscopy can be used to

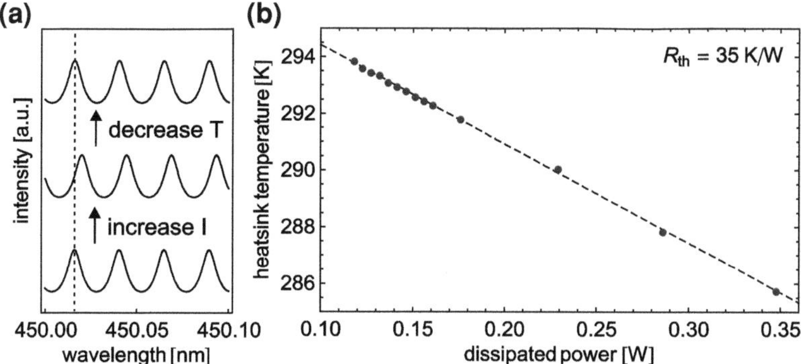

Fig. 3.4 Illustration of the longitudinal mode shift above threshold when current is increased and heatsink temperature is decreased (**a**) and heatsink temperatures with equal longitudinal mode positions versus dissipated power above threshold (**b**). The *dashed line* is a linear fit to the data points, its slope gives the thermal resistance R_{th}

probe the evolution of the temperature in pulsed operation [7]. For this purpose, a streak camera system is used with a temporal resolution better than 1% of the set time range and about 20 pm spectral resolution, which is sufficient to separate the individual longitudinal modes of the laser spectrum. The time evolution of the laser emission in pulsed operation is analyzed on different time scales from 50 ns to 1 ms, at a heat sink temperature of 23°C. A very low duty cycle of 1/1000 is chosen to avoid residual heating. Figure 3.5 shows the longitudinal mode position and the center wavelength as functions of time on different time scales. Two distinct regimes can be recognized: On a time scale of some ten nanoseconds after the onset of the pulse, the center wavelength shifts by several hundred picometers with a time constant of about $\tau_1 = 6$ ns, although the position of the longitudinal modes remains almost constant (Fig. 3.5a, c). On the larger time scale of 5 µs, the longitudinal modes shift by several ten picometers with a time constant of $\tau_2 = 0.4$ µs, and the center wavelength approximately follows this shift (Fig. 3.5b, d, e).

The large redshift of the center wavelength observed within the first 20 ns of the pulse cannot be attributed to charge carrier dynamics, as these manifest typically in relaxation oscillations that decay within few nanoseconds (compare Sect. 2.4). Mode competition, as described in Ref. [8] can also be ruled out as a possible explanation, as this would cause intensity oscillations of the individual modes. From the observations of a constant initial center wavelength of 443.1 nm and the increase of the total redshift with increasing input power, it becomes clear that this shift is a thermal effect. This seems to contradict the observation of a constant longitudinal mode position, which indicates a constant temperature of the crystal lattice. However, the peak of the gain spectrum depends not only on the properties of the crystal, but also on the condition of the charge carrier plasma, as explained above. Although the plasma couples to the lattice via phonons, it forms a separate heat reservoir, which can heat up much faster than the crystal due to its small spatial extension and low heat

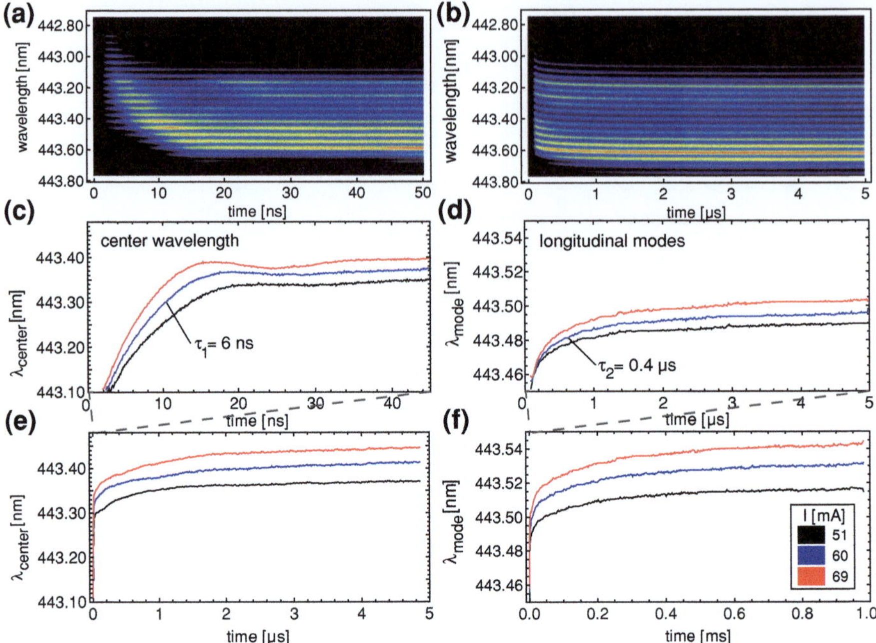

Fig. 3.5 Example Streak camera measurements showing the time evolution of the longitudinal mode spectrum in pulsed operation at 69 mA with 50 ns (**a**) and 5 µs time scale (**b**). Center wavelength as a function of time during the first 50 ns (**c**) and 5 µs (**e**) after the onset of the electrical pulse and position of one longitudinal mode on a 5 µs (**d**) and on a 1 ms time scale (**f**) for three different pump currents

capacity. Therefore, the fast redshift of the center wavelength is due to a rapid self-heating of the charge carrier plasma by intraband carrier relaxation in the quantum wells.

A comparison of the initial value of the center wavelength in pulsed operation to the value in cw operation gives an estimate for the total temperature change of the charge carrier plasma. For a pump current of 60 mA and a heat sink temperature of 23°C, the equilibrium center wavelength is 443.67 nm. This gives a temperature change of the plasma of

$$\Delta T_{\mathrm{plasma}} = (\lambda_{\mathrm{center,final}} - \lambda_{\mathrm{center,initial}}) / \frac{\mathrm{d}\lambda_{\mathrm{center}}}{\mathrm{d}T}$$

$$= (570\,\mathrm{pm})/(36\,\mathrm{pm/K}) = 16\,\mathrm{K}. \tag{3.5}$$

The temperature change of the whole chip is obtained by multiplying the thermal resistance with the input power minus the optical output power P_{opt}:

$$\Delta T_{\mathrm{chip}} = R_{\mathrm{th}}(U \cdot I - P_{\mathrm{opt}})$$

$$= 35\,\mathrm{K/W} \cdot (60\,\mathrm{mA} \cdot 4.9\,\mathrm{V} - 18\,\mathrm{mW}) = 9.7\,\mathrm{K}. \tag{3.6}$$

This result implies that the charge carrier plasma temperature in cw operation is not identical to the temperature of the surrounding crystal lattice. Such a behavior can be expected for coupled subsystems with different heat capacities, like in this case the charge carrier plasma and the crystal lattice.

The heating of the crystal lattice, which is probed by the position of the longitudinal modes, takes place on a much longer time scale. At 60 mA, a shift of the longitudinal modes by about 30 pm during the first 2 µs is observed, which corresponds to a temperature change of the crystal lattice of only about 2 K (compare Sect. 3.1). However, a look at the even longer time scale of 1 ms (see Fig. 3.5f) reveals another contribution to the self-heating, which is much slower and can be attributed to the heating of the submount. Such a behavior was also observed in a study of the thermal properties of GaN-based LEDs using the forward voltage method [9].

To understand the different behaviors of plasma and crystal lattice, the heat capacities of these two subsystems are estimated. The threshold carrier density of highly optimized (Al,In)GaN laser diodes is on the order of $n_{\mathrm{th}} \approx 10^{13} \mathrm{cm}^{-2}$ and the waveguide area is typically $A_{\mathrm{wg}} = 2 \times 650\,\mu\mathrm{m}^2$. The heat capacity of the charge carrier plasma is then

$$c_{\mathrm{plasma}} = 2 \cdot (2k_{\mathrm{B}})n_{\mathrm{th}} A_{\mathrm{wg}} \approx 8 \cdot 10^{-15} \mathrm{J/K}, \qquad (3.7)$$

with the Boltzmann constant k_B. Here, the first factor 2 accounts for electrons and holes, and the heat capacity per particle of the two-dimensional Fermi gas is $2 \cdot \frac{1}{2}k_B$ for the two degrees of freedom plus k_B for the condition of constant pressure.

Using the literature values for the molar volume $V_{\mathrm{c}} = 13.61 \mathrm{cm}^3\mathrm{mol}^{-1}$ and the specific heat at 300 K, $C_{\mathrm{p}} = 9.7\,\mathrm{J/(mol\ K)}$ of GaN [10] and assuming a substrate thickness of $d = 80\,\mu\mathrm{m}$ and chip dimensions of $A_{\mathrm{chip}} = 400 \times 650\,\mu\mathrm{m}^2$, an estimate for the heat capacity of the whole chip is obtained:

$$c_{\mathrm{chip}} = A_{\mathrm{chip}}d \cdot C_{\mathrm{p}}/V_{\mathrm{c}} \approx 1.5 \cdot 10^{-5}\,\mathrm{J/K}. \qquad (3.8)$$

Multiplying this value with the thermal resistance of the chip gives

$$\tau_2 = R_{\mathrm{th}}c_{\mathrm{chip}} = 0.5\,\mu\mathrm{s}, \qquad (3.9)$$

in good agreement with the measured value of τ_2. Plugging the measured τ_1 and the calculated c_{plasma} into Eq. 3.9, gives an estimate for the thermal resistance between plasma and crystal of 8×10^5 K/W. With these values in mind it becomes clear why the charge carrier plasma can heat up so much faster than the crystal despite the fact that almost the entire dissipated energy flows directly into the crystal via ohmic heating, charge carrier leakage and nonradiative recombination. Charge carrier relaxation in the quantum wells also mainly heats the crystal due to the strong electron-phonon coupling and only a small part of this energy goes into the plasma via electron-electron interaction.

As a consequence of the fast timescale of self-heating of the charge carrier plasma, limitations of the output power due to thermal roll-over occur also in pulsed operation at low duty cycle, at a higher pump current than in cw operation. Such limitations in short-pulse operation have also been observed on infrared laser diodes in other material systems [11].

References

1. Nichia Corporation, Data sheet NDHB510APA, http://www.nichia.co.jp/specification/en/product/ld/NDHB510APA-E.pdf
2. L.A. Coldren, S.W. Corzine, *Diode Lasers and Photonic Integrated Circuits* (Wiley, New York, 1995)
3. C. Eichler, S. Schad, M. Seyboth, F. Habel, M. Scherer, S. Miller, A. Weimar, A. Lell, V. Härle, D. Hofstetter, Time resolved study of laser diode characteristics during pulsed operation. Physica Status Solidi C **0**(7), 2283 (2003)
4. G.M. Laws, E.C. Larkins, I. Harrison, C. Molloy, D. Somerford, Improved refractive index formulas for the Al. J. Appl. Phys. **89**(2), 1108–1115 (2001)
5. Y.P. Varshni, Temperature dependence of the energy gap in semiconductors. Physica **34**(1), 149 (1967)
6. B.W. Hakki, T.L. Paoli, cw degradation at 300 K of GaAs double-heterostructure junction lasers. II. Electronic gain. J. Appl. Phys. **44**(9), 4113 (1973)
7. W.G. Scheibenzuber, U.T. Schwarz, Fast self-heating in GaN-based laser diodes. Appl. Phys. Lett. **98**, 181110 (2011)
8. B. Schmidtke, H. Braun, U.T. Schwarz, D. Queren, M. Schillgalies, S. Lutgen, U. Strauss, Time resolved measurement of longitudinal mode competition in 405 nm (Al, In)GaN laser diodes. Physica Status Solidi C **6**, S860–S863 (2009)
9. Q. Shan, Q. Dai, S. Chhajed, J. Cho, E.F. Schubert, Analysis of thermal properties of GaInN light-emitting diodes and laser diodes. Journal **108**, 084504 (2010)
10. H. Morkoc, *Handbook of Nitride Semiconductors and Devices: Materials Properties 1, Physics and Growth* (Wiley-VCH, Weinheim, 2008)
11. M. Mueller, M. Rattunde, G. Kaufel, J. Schmitz, J. Wagner, Short-pulse high-power operation of GaSb-based diode lasers. IEEE Photonics Technol. Lett. **21**(23), 1770–1772 (2009)

Chapter 4
Light Propagation and Amplification in Laser Diodes from Violet to Green

GaN-based laser diodes can be fabricated for emission wavelengths within a wide spectral range, from near-ultraviolet to green. While highly efficient violet and blue laser diodes are available as commercial products, green laser diodes still suffer from high threshold currents and low output power. The worse performance of green laser diodes originates from the high indium content in the quantum wells, which is necessary to achieve green emission.

In this chapter, the propagation and amplification of light in GaN-based laser diodes emitting at different wavelengths from violet to green are compared with respect to their impact on the output characteristics of the devices. Optical gain and refractive index spectra are analyzed at different pump current levels, and the different properties are related to the indium content of the devices. From the changes in refractive index and gain with current, the antiguiding factor is calculated, an empirical quantity that is used to describe dynamical phenomena in laser diodes.

The samples studied in this chapter are prototype laser diodes grown on free-standing GaN substrates at *OSRAM Opto Semiconductors* [1, 2, 3] by MOVPE. They have a multi-quantum-well active region with thin InGaN quantum wells (QW thickness <4 nm). Their emission wavelengths are 409 nm (violet), 457 nm (blue) and 511 nm (green).

4.1 Output Characteristics

Especially for mobile, battery-powered applications, a high efficiency of the laser diodes is necessary. This requires a low threshold current, a high slope efficiency and a low forward voltage. Figure 4.1 shows the output characteristics of the green, blue and violet laser diode. Their threshold currents and forward voltages at threshold are 36, 24, 104 mA, and 4.9, 4.2, 6.4 V, respectively. While the slope efficiency has values of 1.30 and 1.06 W/A for the violet and blue laser diode, it is only 0.36 W/A for the green laser diode.

W. G. Scheibenzuber, *GaN-Based Laser Diodes*, Springer Theses,
DOI: 10.1007/978-3-642-24538-1_4, © Springer-Verlag Berlin Heidelberg 2012

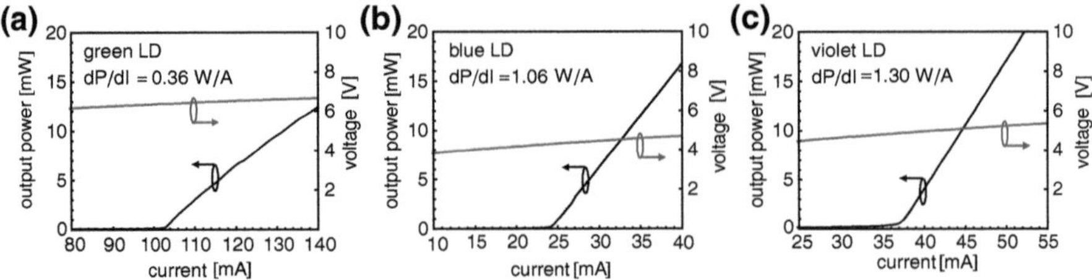

Fig. 4.1 Forward voltage and output power as functions of current for the green (**a**), blue (**b**) and violet (**c**) laser diode. The current scales are different for each graph

The threshold current of the green laser diode is three to four times higher than for the violet and blue laser diodes. As it is shown in the following section, the high threshold current of the green laser diode is a consequence of drastically reduced optical gain per current, compared to the blue and violet LD. At such a high current, self-heating effects become relevant, increasing the threshold and leading to a thermal roll-over of the output power curve. At 10 mW optical power, the overall efficiency ("wall-plug" efficiency), that is the quotient of output and input power, is 0.01, 0.08 and 0.05 for the green, blue and violet LD, respectively.

4.2 Optical Gain

A high threshold current can be caused either by high optical losses, low injection efficiency or insufficient optical gain. To differentiate these effects, optical gain spectra are measured using the Hakki-Paoli method [4]. The method is based on a high resolution measurement of the amplified spontaneous emission (ASE) spectra below threshold. For that purpose, a Czerny–Turner double spectrometer of 0.85 m focal length with 1,800 lines/mm gratings with a spectral resolution better than 4 pm is used. Figure 4.2 shows an example spectrum of the violet laser diode at an intermediate pump current. The spectrum shows regular modulations which correspond to the longitudinal modes of the (Fabry–Perot) laser resonator. Amplitude and spectral position of these modulations are determined by the refractive index and the overall optical loss in the resonator. The net absorption, which is given by the internal and mirror losses minus the gain (compare Sect. 2.3), is calculated from the modulation depth $P_{\max}/P_{\min}$,

$$\alpha(\hbar\omega) = \frac{1}{L} \ln\left(\frac{\sqrt{P_{\max}/P_{\min}} + 1}{\sqrt{P_{\max}/P_{\min}} - 1}\right), \tag{4.1}$$

where $P_{\max}$ and $P_{\min}$ are the intensities of adjacent maxima and minima of the ASE spectrum and L is the cavity length of the laser diode, which is 650 μm for the three samples studied in this chapter.

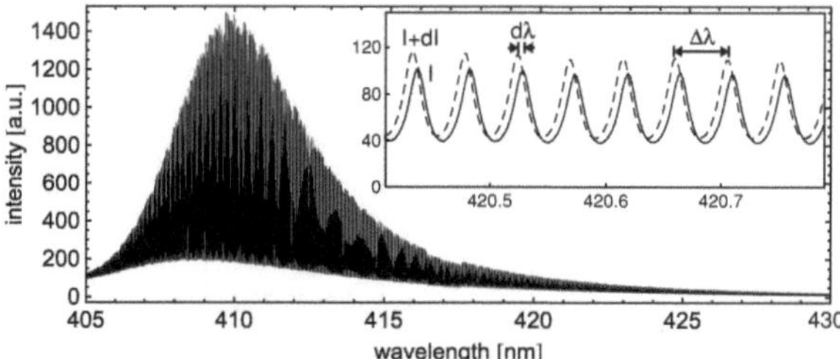

Fig. 4.2 High-resolution amplified spontaneous emission spectrum of the violet laser diode at 25 mA. The inset shows a magnification of a part of the spectrum at 25 mA (*solid*) and 30 mA (*dashed*)

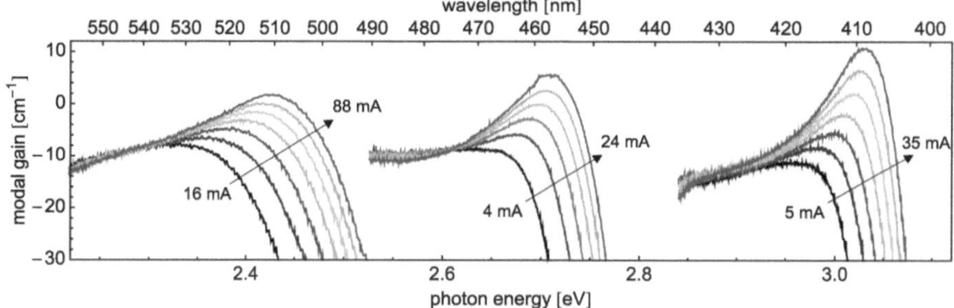

Fig. 4.3 Optical gain spectra of the green, blue and violet laser diodes (from *left* to *right*). The pump currents vary from 16 to 88mA in steps of 12 mA, 4 − 24 mA in steps of 4 mA, and 5 − 35 mA in steps of 5 mA for the green, blue and violet laser diode, respectively (*bottom* to *top*)

Optical gain spectra of the green, blue and violet laser diode are investigated at various pump currents below threshold. The optical gain depends not only on the pump current, but also on the internal temperature, which increases with increasing current due to self-heating. In order to avoid the thermal influence, the thermal resistances of the laser diodes are measured using the method described in Sect. 3.2. where values of 33, 21 and 26 K/W are obtained for the green, blue and violet LD, respectively. The heat sink, to which each diode is mounted during the measurement, is then cooled down successively with increasing pump current, so the internal device temperature is kept constant.

Figure 4.3 shows optical gain spectra of the three samples, which are shifted vertically by the mirror losses of each laser diode. The internal losses can then be determined from the long wavelength tail of the spectra, where there is no gain or absorption from the active region and the curves for all currents converge. The green, blue and violet laser diodes have low internal absorption losses of 9, 9 and 12cm^{-1}, respectively. For the green laser diode, the internal losses show a slight wavelength dependency due to a reduction of the optical mode confinement with increasing wavelength. Using Eq. 2.19 and the measured slope efficiencies, the injection efficiencies are determined as 0.62 (green), 1.0 (blue) and 0.80 (violet).

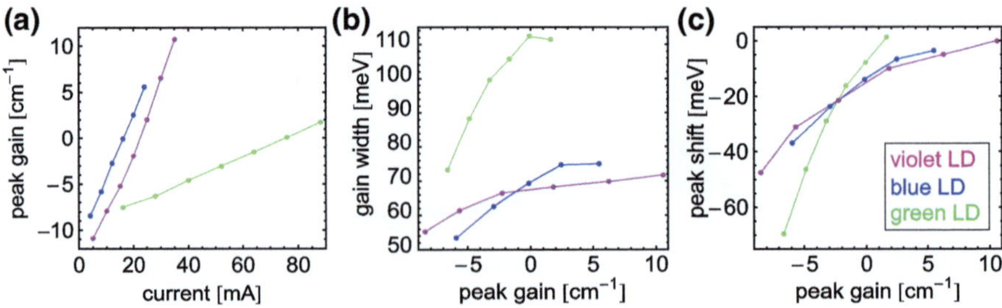

Fig. 4.4 Peak gain as a function of current (**a**), spectral width (full width half maximum, FWHM) of the gain curve (**b**) and energetic shift of peak position relative to the laser emission energy (**c**) as functions of peak gain for the green, blue and violet laser diode

Table 4.1 Summary of device parameters for the green, blue and violet laser diode

Parameters	Green LD	Blue LD	Violet LD
Internal losses α_{int} [cm^{-1}]	9	9	12
Injection efficiency η_{inj}	0.62	1.0	0.80
Differential peak gain dg/dI [cm^{-1} mA^{-1}]	0.13	0.70	0.86
FWHM of gain spectrum at threshold [meV]	112	75	72

Towards longer laser wavelength, the gain spectra show an increase of spectral width and a decrease of differential peak gain dg/dI (see Fig. 4.4a, b). Furthermore, a blue-shift of the gain peak with increasing pump current is observed (see Fig. 4.4c), which becomes stronger from violet to green. These effects are related to the quantum confined Stark effect and the reduction of the wave function overlap with increasing indium content in the quantum wells due to the internal field (compare Sect. 2.2.2). The reduced wave function overlap causes a reduction of the optical gain and consequently a higher threshold charge carrier density. With increasing charge carrier density, the Fermi–Dirac distribution of the charge carriers causes an increasing occupation of higher-energetic states in the band structure, broadening the spectra and shifting the maximum to higher energy. The spatial distribution of charge carriers in the quantum wells partially screens the internal fields and causes an additional blue-shift of the emission. Another contribution to the spectral width of the gain spectra is the increase of the inhomogeneous broadening with increasing indium content due to a reduction of the crystal quality [5]. Particularly for green laser diodes, the strong coupling of charge carriers to longitudinal optical phonons also affects the spectral shape of the gain spectra [6]. Table 4.1 summarizes the extracted gain parameters.

4.3 Refractive Index

From the longitudinal modes in the ASE spectrum, not only the optical gain can be deduced, but also the real part of the complex susceptibility, that is the refractive index. Real and complex part are related to each other via the Kramers–Kronig-Relation. Therefore, a change in optical gain is accompanied by a charge carrier induced change of refractive index. Figure 4.5 shows a small part of the spectrum of the violet laser diode for different pump currents at constant internal temperature. With increasing current, the longitudinal modes become sharper due to the increase of the optical gain, and their positions shift to blue due to the change of the refractive index. The spacing $\Delta\lambda$ of the longitudinal modes is given by the group refractive index

$$\Delta\lambda = \frac{\lambda^2}{2Ln_{\mathrm{gr}}}, \tag{4.2}$$

$$n_{\mathrm{gr}} = n_{\mathrm{eff}} - \lambda\frac{\mathrm{d}n_{\mathrm{eff}}}{\mathrm{d}\lambda}. \tag{4.3}$$

Group refractive index spectra (see Fig. 4.6) can thus be extracted from the ASE measurements by evaluating the mode spacing as a function of photon energy. The refractive indices of the waveguide materials GaN, InGaN and AlGaN decrease with increasing wavelength, so this tendency also applies to the group refractive index. At the laser wavelength, it has values of 2.75 (green), 2.90 (blue) and 3.27 (violet). The group refractive index is a required parameter for dynamical simulations of laser diodes, as it determines the speed of light in the resonator (compare Sect. 2.4). The charge carrier induced refractive index change $\mathrm{d}n/\mathrm{d}I$ relates to the change of the longitudinal mode position $\mathrm{d}\lambda/\mathrm{d}I$ (compare Fig. 4.2) [7],

$$\frac{\mathrm{d}\lambda}{\mathrm{d}I} = \frac{\lambda}{n_{\mathrm{gr}}}\frac{\mathrm{d}n}{\mathrm{d}I}. \tag{4.4}$$

Integrating the spectra of $\mathrm{d}n/\mathrm{d}I$ obtained from the ASE data yields the total modal refractive index change Δn, as shown in Fig. 4.7. The refractive index change spectra show a similar tendency as the optical gain spectra: While for the violet LD, they exhibit a sharp slope to shorter wavelength, they appear rather flat for the green LD, which is a consequence of the increased inhomogeneous broadening.

As the internal temperature is kept constant, the optical properties of the waveguide layers do not change. There is thus no contribution of the temperature dependence of the refractive index $\mathrm{d}n/\mathrm{d}T$ to the refractive index change Δn (Fig. 4.7). Modal gain g and modal refractive index change Δn are given by the material gain and refractive index change of the InGaN quantum wells, multiplied with the optical confinement factor Γ.

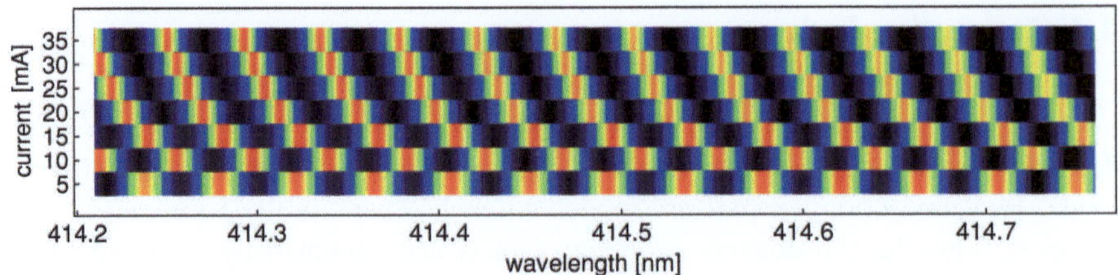

Fig. 4.5 Emission intensity (*color-coded*) below threshold as a function of wavelength and pump current

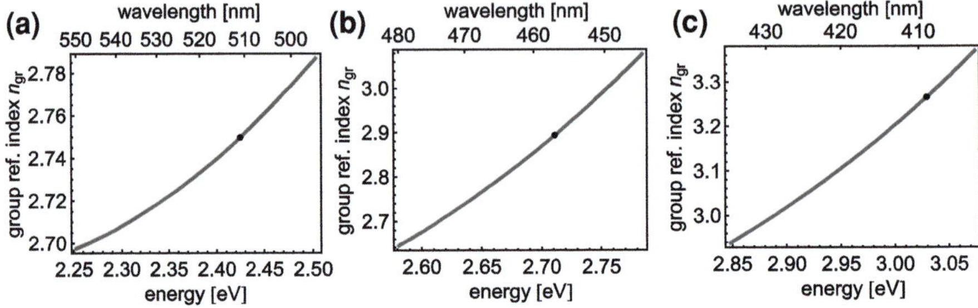

Fig. 4.6 Group refractive index spectra of the green (**a**), blue (**b**) and violet (**c**) laser diode. The *black points* indicate the laser energies of the respective samples

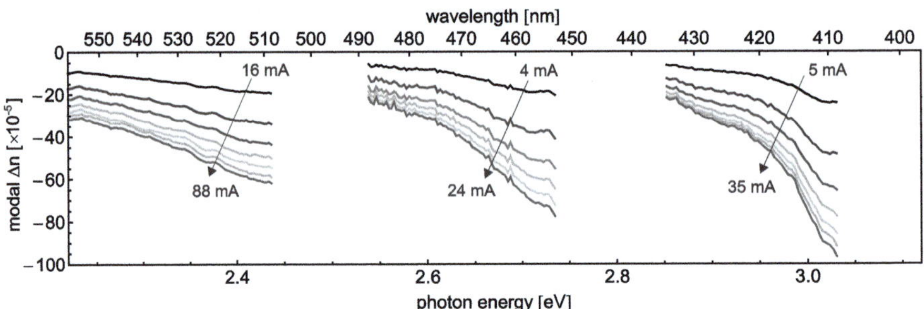

Fig. 4.7 Refractive index change spectra of the green, blue and violet laser diodes (from *left* to *right*). The pump currents vary from 16 to 88 mA in steps of 12 mA, 4–24 mA in steps of 4 mA, 5–35 mA in steps of 5 mA for the green, blue and violet laser diode, respectively (*top* to *bottom*)

4.4 Antiguiding Factor

The antiguiding factor is a simplified empirical concept to describe the interplay of optical gain and refractive index change in a laser diode. It is relevant for the design of single-mode external cavity lasers, as it determines the laser linewidth [8] and the stability of longitudinal single-mode operation [9, 10]. Furthermore it is important for high-power broad area laser diodes, as a high antiguiding factor means an increased tendency of the laser to the formation of higher order lateral modes [11] and the

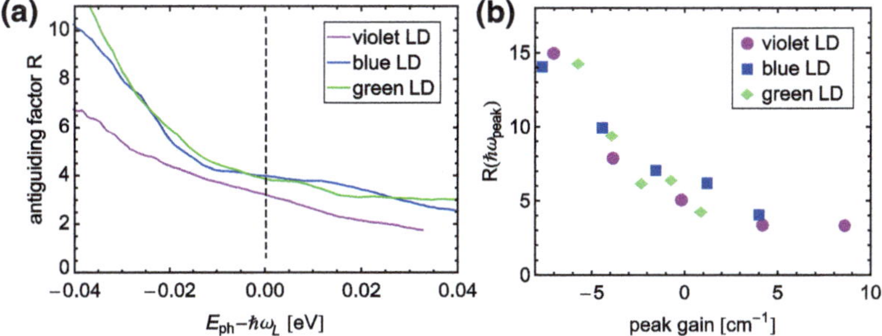

Fig. 4.8 Antiguiding factor slightly below threshold as function of photon energy relative to laser emission energy (**a**) and antiguiding factor at gain peak energy as function of peak gain (**b**)

occurrence of filaments [10, 12]. It is also a required parameter for simulations of laser dynamics [9], especially for short pulse laser diodes.

The antiguiding factor is defined as the refractive index change per gain change,

$$R = -\frac{4\pi}{\lambda}\frac{dn/dI}{dg/dI}.$$
(4.5)

It is not influenced by the optical mode confinement, as Γ drops out in Eq. 4.5. For all samples, the antiguiding factor has a value of about 3 to 4 at the laser wavelength. This finding is in good agreement with earlier measurements on violet laser diodes [7] and theoretical predictions for narrow quantum wells [11]. The values of R at the laser wavelengths for the highest currents are 4.3 (green), 4.1 (blue) and 3.4 (violet), with a systematic error of ± 0.5. Figure 4.8 shows the spectral dependency of the antiguiding factor and its value at the peak wavelength as a function of peak gain for all three samples. While the antiguiding factor can be as high as 15 at low gain, it goes down to values below five above the transparency threshold for all three emission wavelengths. The observation of a similar antiguiding factor for all three samples indicates that the antiguiding factor is not significantly influenced by the quantum confined Stark effect. Although the gain spectra and the refractive index change spectra are strongly influenced by the QCSE, this influence apparently almost cancels out for the antiguiding factor. The strong dependence of the antiguiding factor on the QW width, which is reported in Ref. [11], may thus rather be caused by changes in the shape of the gain curve due to contributions from higher energetic states in the quantum wells.

References

1. S. Brueninghoff, C. Eichler, S. Tautz, A. Lell, M. Sabathil, S. Lutgen, U. Strauss, 8 W single-emitter InGaN laser in pulsed operation. Physica Status Solidi A **206**, 1149 (2009)
2. U. Strauss, C. Eichler, C. Rumbolz, A. Lell, S. Lutgen, S. Tautz, M. Schillgalies, S. Brüninghoff, Beam quality of blue InGaN laser for projection. Physica Status Solidi C **5**, 2077 (2008)

3. A. Avramescu, T. Lermer, J. Müller, S. Tautz, D. Queren, S. Lutgen, U. Strauss, InGaN laser diodes with 50 mW output power emitting at 515 nm. Appl. Phys. Lett. **95**, 071103 (2009)
4. B.W. Hakki, T.L. Paoli, cw degradation at 300 K of GaAs double-heterostructure junction lasers. II. Electronic gain. J. Appl. Phys. **44**(9), 4113 (1973)
5. K. Kojima, U.T. Schwarz, M. Funato, Y. Kawakami, S. Nagahama, T. Mukai, Optical gain spectra for near UV to aquamarine (Al,In) GaN laser diodes. Optics Express **15**, 7730–7736 (2007)
6. T. Lermer, A. Gomez-Iglesias, M. Sabathil, J. Müller, S. Lutgen, U. Strauss, B. Pasenow, J. Hader, J.V. Moloney, S.W. Koch, W. Scheibenzuber, U.T. Schwarz, Gain of blue and cyan InGaN laser diodes. Appl. Phys. Lett. **98**(2), 021115 (2011)
7. U.T. Schwarz, E. Sturm, W. Wegscheider, V. Kümmler, A. Lell, V. Härle, Optical gain, carrier-induced phase shift, and linewidth enhancement factor in InGaN quantum well lasers. Appl. Phys. Lett. **83**(20), 4095 (2003)
8. C. Henry, Theory of the linewidth of semiconductor lasers. IEEE J. Quantum Electron. **18**, 259 (1982)
9. W.W. Chow, S.W. Koch, *Semiconductor-Laser Fundamentals* (Springer, Berlin, 1998)
10. M. Osinski, J. Buus, Linewidth broadening factor in semiconductor lasers—an overview. IEEE J. Quantum Electron. **23**(1), 9 (1987)
11. W.W. Chow, H. Amano, I. Akasaki, Theoretical analysis of filamentation and fundamental-mode operation in InGaN quantum well lasers. Appl. Phys. Lett. **76**(13), 1647–1649 (2000)
12. D. Scholz, H. Braun, U.T. Schwarz, A. Lell, U. Strauss, Measurement and simulation of filamentation in (Al,In)GaN laser diodes. Optics Express **16**(10), 349–353 (2008)

Chapter 5
Semipolar Crystal Orientations for Green Laser Diodes

Although there has been an enormous progress in the development of GaN-based true-green laser diodes on c-plane substrates [1, 2], these devices still have significantly lower efficiency and higher threshold currents than blue and violet laser diodes. The worse performance of GaN-based green laser diodes relates to a reduction and broadening of the optical gain (compare Chap. 4), which is caused by the lower material quality of epitaxial layers with high indium content and the reduction of the electron-hole wave function overlap in the quantum wells due to the internal piezoelectric fields. While the material quality can be improved to a certain degree by optimizing the epitaxial growth, the reduction of the wave function overlap poses a fundamental limitation for quantum well devices grown on c-plane GaN. Hence it is doubtful whether further improvements in growth conditions and heterostructure design can enable the fabrication of green laser diodes with an output power of several hundred mW and a sufficiently long device lifetime.

A way to avoid this limitation is the growth on other crystal planes than the c-plane. In the wurzite crystal, the spontaneous and piezoelectric polarizations are oriented along the c-direction. Growing a heterostructure on a crystal plane which is inclined towards the c-direction reduces the polarization discontinuities along the growth direction and thus also the internal fields. Owing to the reduction or extinction of polarization fields, these crystal planes have been termed semipolar or nonpolar. Figure 5.1 shows the orientation of various polar, semipolar and nonpolar crystal planes and their inclination angle to the c-plane. Continuous wave blue and green laser diodes have already been demonstrated on these crystal planes [3, 4]. These devices still do not show better performance than the best c-plane laser diodes, but crystal quality and heterostructure design in semipolar laser diodes are by far not as optimized as they are in c-plane devices. Therefore, vast improvements can be expected for green laser diodes on semipolar GaN.

To reach green laser emission on semi- or nonpolar planes, a higher indium content in the quantum wells is required than on the c-plane, owing to the smaller redshift of the emission due to the reduced quantum-confined Stark effect. Although the nonpolar planes have the advantage of vanishing internal fields, they suffer from limi-

W. G. Scheibenzuber, *GaN-Based Laser Diodes*, Springer Theses,
DOI: 10.1007/978-3-642-24538-1_5, © Springer-Verlag Berlin Heidelberg 2012

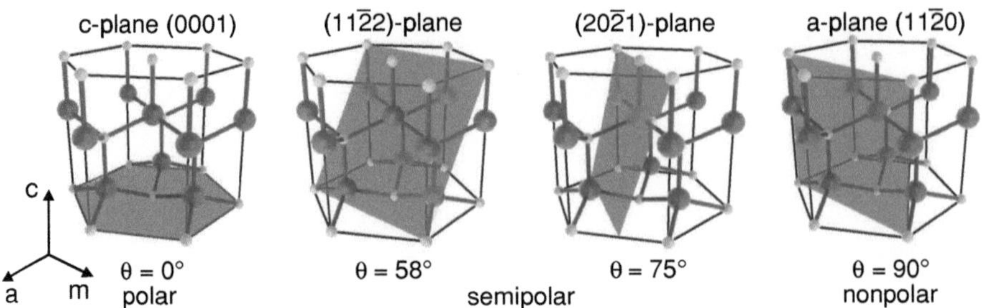

Fig. 5.1 Orientations of c-plane, semipolar (11$\bar{2}$2)- and (20$\bar{2}$1)-planes and nonpolar (11$\bar{2}$0)-plane (*shaded polygons*) with inclination angle θ to the c-plane

tations of the indium incorporation during growth [5, 6], which make them unsuitable for green laser diodes. On the other hand, the indium incorporation on the semipolar (11$\bar{2}$2)- and (20$\bar{2}$1)-planes is similar to the c-plane and the internal fields are drastically reduced on these semipolar planes. Therefore, these two planes are promising candidates for the realization of high power green laser diodes.

This chapter describes the improvements in wave function overlap and optical gain for laser diodes on semipolar crystal planes compared to c-plane devices. In semipolar devices, anisotropy effects on the band structure and the optical polarization of waveguide modes arise from the tilted crystal orientation. These effects are discussed with respect to their consequences on the optical gain. Furthermore, the phenomenon of polarization switching, which occurs on the (11$\bar{2}$2)-plane is explained. The material parameters employed for the calculations in this chapter are taken from Ref. [7], if not stated otherwise, and the numerical methods are described in appendix A.

Two coordinate systems are used to describe non-c-plane devices: The crystal coordinate system x, y, z, where z points along the c-direction and x, y lie in the c-plane, and the growth coordinate system x', y', z', where z' is perpendicular to the growth plane (see Fig. 5.2). In the notation presented here, $y' = y$ is the in-plane direction perpendicular to the c-axis and x' is the projection of the c-axis on the growth plane (c'-direction). For the (11$\bar{2}$2)-plane, y' corresponds to the m-direction of the crystal, whereas for the (20$\bar{2}$1)-plane y' is the a-direction. The two coordinate systems are related via a polar rotation around the y-axis by the crystal angle θ:

$$\begin{pmatrix} x' \\ y' \\ z' \end{pmatrix} = U \begin{pmatrix} x \\ y \\ z \end{pmatrix} \tag{5.1}$$

$$U = \begin{pmatrix} \cos(\theta) & 0 & -\sin(\theta) \\ 0 & 1 & 0 \\ \sin(\theta) & 0 & \cos(\theta) \end{pmatrix} \tag{5.2}$$

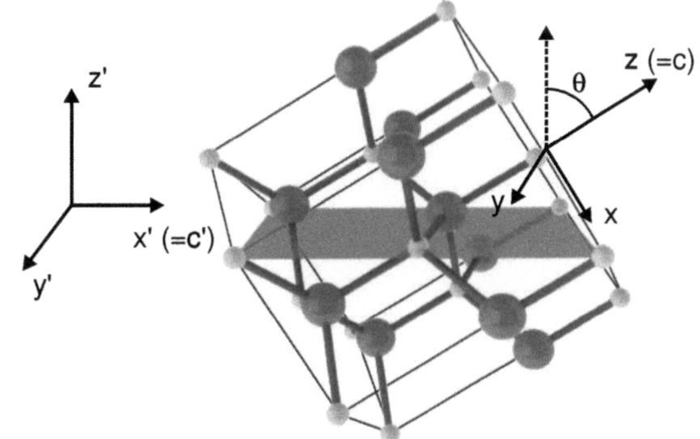

Fig. 5.2 Orientation of crystal coordinate system x, y, z and growth coordinate system x', y', z'. The *shaded polygon* marks the $(1\bar{1}2\bar{2})$-plane

5.1 Reduction of Internal Electric Field

As stated above, the main purpose of growth on crystal planes with a different orientation than the commonly used c-plane is the reduction of the internal electric fields. The dominant contribution to the internal fields comes from the piezoelectric polarization of the quantum wells, which is a function of the strain of these layers. The strain tensor ϵ depends on the inclination angle θ (crystal angle) of the growth plane to the c-plane [8],

$$\epsilon_{xx} = \epsilon_{xx}^{(0)} + \epsilon_{xz}\tan\theta, \tag{5.3}$$

$$\epsilon_{yy} = \epsilon_{xx}^{(0)}, \tag{5.4}$$

$$\epsilon_{zz} = \epsilon_{zz}^{(0)} + \epsilon_{xz}\cot\theta, \tag{5.5}$$

$$\epsilon_{xz} = -\frac{\left((c_{11} + c_{12} + c_{13}\frac{\epsilon_{zz}^{(0)}}{\epsilon_{xx}^{(0)}})\sin^2\theta + (2c_{13} + c_{33}\frac{\epsilon_{zz}^{(0)}}{\epsilon_{xx}^{(0)}})\cos^2\theta\right)\epsilon_{xx}^{(0)}\cos\theta\sin\theta}{c_{11}\sin^4\theta + 2(c_{13} + 2c_{44})\sin^2\theta\cos^2\theta + c_{33}\cos^4\theta}, \tag{5.6}$$

with $\epsilon_{xx}^{(0)} = (a_s - a_e)/a_e$ and $\epsilon_{zz}^{(0)} = (c_s - c_e)/c_e$. The other strain coefficients are zero. a_s, c_s, a_e, c_e are the lattice constants of the substrate and the quantum well, respectively and c_{ij} are the elastic constants of the strained layer. The coordinates x, y, z correspond to the crystal directions a, m, c, respectively (see Fig. 5.1). Figure 5.3 shows the components of the strain tensor as functions of the crystal angle. The polarization discontinuity $\Delta P_{z'}$ along the growth direction z' can then be calculated from the strain tensor [8]:

$$\Delta P_z = \left([d_{31}(c_{11} + c_{12}) + d_{33}c_{13}](\epsilon_{xx} + \epsilon_{yy}) + (2d_{31}c_{13} + d_{33}c_{33})\epsilon_{zz} \right)\cos\theta$$

$$+ (2d_{15}c_{44}\epsilon_{xz})\sin\theta + \left(P_{\text{sp}}^{(\text{QW})} - P_{\text{sp}}^{(\text{b})}\right)\cos\theta, \tag{5.7}$$

with the spontaneous polarization $P_{\text{sp}}^{(\text{QW/b})}$ in the quantum well and barrier and the piezoelectric coefficients d_{ij}, which are taken from Ref. [9]. The internal electric

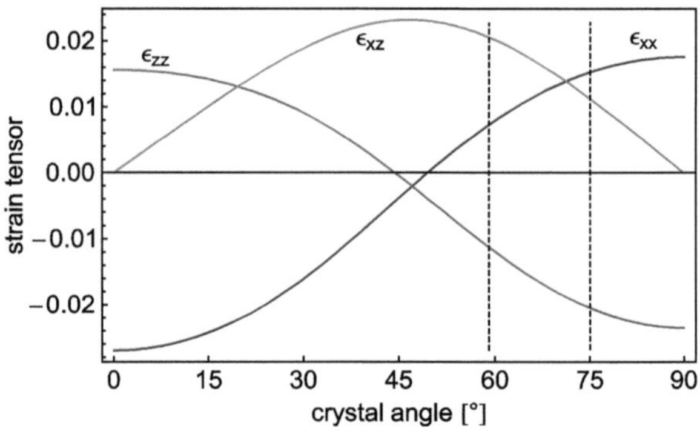

Fig. 5.3 Strain tensor components as function of crystal angle for a fully strained $In_{0.25}Ga_{0.75}N$ layer grown on free-standing GaN substrate. The *dashed lines* mark the semipolar $(11\bar{2}2)$- and the $(20\bar{2}1)$-planes

Fig. 5.4 Polarization discontinuity along growth direction and wave function overlap as functions of crystal angle for a 3 nm wide $In_{0.25}Ga_{0.75}N$ quantum well. The *dashed lines* mark the semipolar $(11\bar{2}2)$- and the $(20\bar{2}1)$-planes

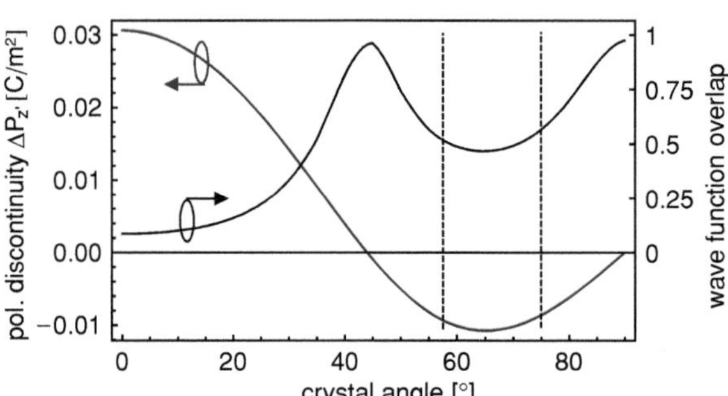

fields in the barrier and QW can then be estimated by assuming appropriate boundary conditions and applying the Gauss law (Eq. 2.2) [10]:

$$\varepsilon^{(QW)} E_z^{(QW)} - \varepsilon^{(b)} E_z^{(b)} = -\frac{\Delta P_{z'}}{\varepsilon_0}. \tag{5.8}$$

Here, ε_0 is the permittivity of vacuum, $\varepsilon^{(b)}$ and $\varepsilon^{(QW)}$ are the static dielectric constants for the barrier layers and the quantum well, respectively. For a more accurate calculation of the internal fields, a drift-diffusion simulation has to be employed, which takes into account the field of the p-n junction, the externally applied bias voltage and the current distribution in the device.

Figure 5.4 shows the polarization discontinuity and the resulting wave function overlap as functions of crystal angle for a 3 nm quantum well with an indium content of 25%. The polarization discontinuity changes its sign at a crystal angle of about 45° and becomes zero again at 90°. At these points, the internal field vanishes and consequently the wave function overlap is almost 1. In between these two angles, the polarization discontinuity is drastically reduced with respect to the c-plane. The wave function overlap is increased from 8% for c-plane to about 60% for the semipolar $(11\bar{2}2)$- and $(20\bar{2}1)$-planes. This effect is even more pronounced for higher indium contents.

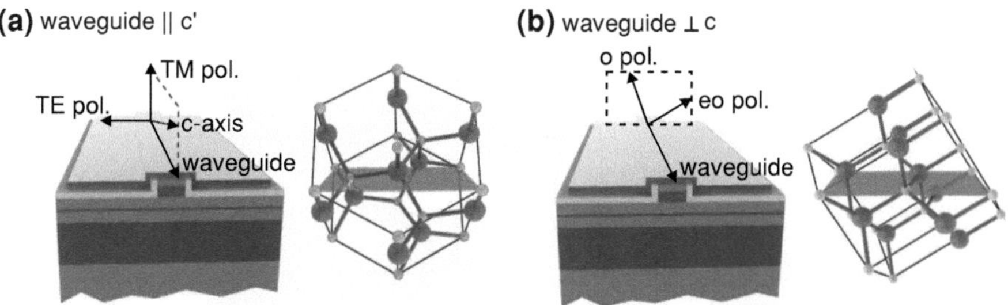

Fig. 5.5 Ridge waveguide orientations relative to the wurzite unit cell for a laser diode grown on the $(11\bar{2}2)$-plane. **a** waveguide parallel to the c'-direction, **b** waveguide perpendicular to the c-direction

5.2 Birefringence in Semipolar Waveguides

Wurzite (Al,In)GaN alloys show uniaxial birefringence with $\Delta n = n_e - n_o \approx 0.01 n_o$ [11], with the extraordinary direction pointing along the c-axis. The $\lambda/2$-distance of a-plane GaN for a wavelength of 510 nm is about 9 μm, which is just a fraction of the length of a typical laser diode waveguide. Therefore, the effect of birefringence has to be taken into account in calculations of the eigenmodes of the waveguide.

In a c-plane laser diode, the extraordinary direction is perpendicular to the growth plane. The eigenmodes of the electric field in these diodes are called transversal electric (TE) or transversal magnetic (TM). Their optical polarization is perpendicular or parallel to the c-axis, respectively. If a laser diode structure is grown on a crystal plane which is inclined to the c-plane, and a ridge waveguide is etched perpendicular to the extraordinary direction, then there is a competition between the transversal refractive index profile, which favors polarizations parallel or perpendicular to the growth plane and the optical anisotropy, which draws the polarization of the modes into the ordinary and extraordinary directions intrinsic to the crystal.

Two possible waveguide orientations for a non-c-plane quantum well have to be distinguished: The ridge waveguide can be either parallel or perpendicular to the projection of the c-axis in the QW-plane (c'-direction), as illustrated in Fig. 5.5. For the first possibility, the ordinary and extraordinary directions perpendicular to the propagation direction match the TE- and TM-polarization directions and thus TE- and TM-modes exist. For this orientation, the mirror facet plane is a high index plane, which complicates the cleavage of the facets. The alternative orientation offers a low index mirror facet plane (at least for devices on the $(11\bar{2}2)$-plane), but the extraordinary direction is then rotated in a plane perpendicular to the propagation direction, causing optical modes different from TE and TM, with polarizations parallel and perpendicular to the c-axis. The optical polarization has to be considered when calculating the optical gain, as it determines the direction of the relevant momentum matrix element (compare Eq. 2.6).

To calculate the optical modes in a waveguide perpendicular to the c-axis, a full-vectorial 4×4-transfer matrix method is used [12]. The dielectric tensor ε in the

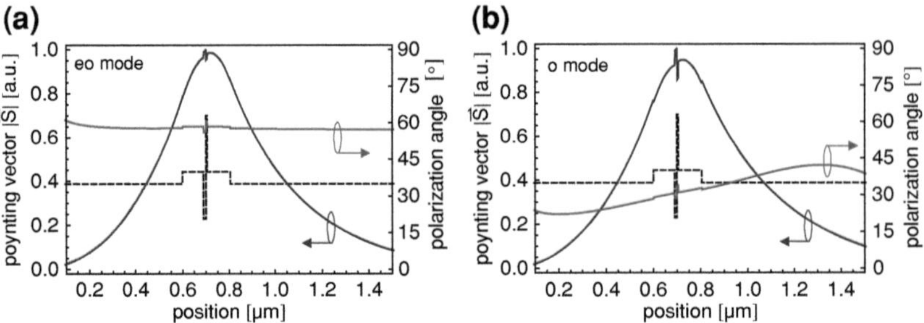

Fig. 5.6 Modulus of the poynting vector and polarization angle distribution of the extraordinary (**a**) and ordinary (**b**) mode of a birefringent laser waveguide on the $(11\bar{2}2)$-plane ($\theta = 58°$). The *dashed line* indicates the refractive index profile

waveguide coordinate system is obtained by rotating the dielectric tensor of a c-plane structure by the crystal angle θ around the propagation direction with the matrix U (Eq. 5.2):

$$\varepsilon = U \begin{pmatrix} n_o^2 & 0 & 0 \\ 0 & n_o^2 & 0 \\ 0 & 0 & n_e^2 \end{pmatrix} U^{\mathrm{T}} \tag{5.9}$$

Waveguide modes are calculated for semipolar single quantum well laser diodes on the $(11\bar{2}2)$-plane at a wavelength of 510 nm. The results for the fundamental mode (the one with the highest effective index of refraction) and the first mode perpendicular to the fundamental one are shown in Fig. 5.6. The average polarization of the eigenmodes points approximately along the ordinary or extraordinary direction of the crystal, therefore these modes are called extraordinary (eo) and ordinary (o) mode. The effective indices of refraction for these two modes are $n_{\mathrm{eff}} = 2.421$ for the extraordinary and $n_{\mathrm{eff}} = 2.397$ for the ordinary mode. In both cases the modes are linearly polarized along the ordinary or extraordinary direction in the waveguide layer with a maximum deviation of about $4°$. The transversal component of the electric field is discontinuous due to the continuity requirement of the dielectric displacement at the material boundaries, where the refractive index changes abruptly. Therefore, the intensity and the polarization angle also exhibit discontinuities at the layer interfaces.

The occurrence of ordinary and extraordinary modes in semipolar GaN waveguides has been predicted in Ref. [10] and was verified experimentally in Refs. [13] and [14].

5.3 Band Structure of Semipolar Quantum Wells

In a semipolar quantum well, the conduction band is almost the same as in a c-plane QW because the electron effective mass is nearly isotropic and the angular momentum eigenfunction of the electrons is the spherically symmetric S-function. On the other hand, the hole band parameters and the strain of wurzite (Al,In)GaN are strongly anisotropic. Therefore, the crystal orientation has a major impact on the valence band structure of an InGaN quantum well. To calculate the valence band structure in a semipolar quantum well, the 6×6 $k \cdot p$ method is used. The following basis is chosen for the hole wave function:

$$|U_1\rangle = -\frac{1}{\sqrt{2}}(|P_x \uparrow\rangle + i|P_y \uparrow\rangle), \ |U_2\rangle = \frac{1}{\sqrt{2}}(|P_x \uparrow\rangle - i|P_y \uparrow\rangle), \ |U_3\rangle = |P_z \uparrow\rangle,$$

$$|U_4\rangle = \frac{1}{\sqrt{2}}(|P_x \downarrow\rangle - i|P_y \downarrow\rangle), \ |U_5\rangle = -\frac{1}{\sqrt{2}}(|P_x \downarrow\rangle + i|P_y \downarrow\rangle), \ |U_6\rangle = |P_z \downarrow\rangle.$$

$$(5.10)$$

For this basis, the valence band Hamiltonian in the crystal coordinate system (x, y, z) for a bulk wurzite crystal is [15]:

$$\mathbf{H}_h(k_x, k_y, k_z) = \begin{pmatrix} F & -K^* & -H^* & 0 & 0 & 0 \\ -K & G & H & 0 & 0 & \Delta \\ -H & H^* & \Lambda & 0 & \Delta & 0 \\ 0 & 0 & 0 & F & -K & H \\ 0 & 0 & \Delta & -K^* & G & -H^* \\ 0 & \Delta & 0 & H^* & -H & \Lambda \end{pmatrix}, \tag{5.11}$$

$$F = \Delta_{\mathrm{cr}} + \frac{\Delta_{\mathrm{so}}}{3} + \Lambda + \Theta,$$

$$G = \Delta_{\mathrm{cr}} - \frac{\Delta_{\mathrm{so}}}{3} + \Lambda + \Theta,$$

$$\Lambda = \frac{\hbar^2}{2m_0}(A_1 k_z^2 + A_2(k_x^2 + k_y^2)) + \Lambda_\epsilon,$$

$$\Theta = \frac{\hbar^2}{2m_0}(A_3 k_z^2 + A_4(k_x^2 + k_y^2)) + \Theta_\epsilon,$$

$$K = \frac{\hbar^2}{2m_0}A_5(k_x + ik_y)^2 + D_5(\epsilon_{xx} - \epsilon_{yy}),$$

$$H = \frac{\hbar^2}{2m_0}A_6(k_x + ik_y)k_z + D_6(\epsilon_{xz} + i\epsilon_{yz}),$$

$$\Lambda_\epsilon = D_1\epsilon_{zz} + D_2(\epsilon_{xx} + \epsilon_{yy}),$$

$$\Theta_\epsilon = D_3\epsilon_{zz} + D_4(\epsilon_{xx} + \epsilon_{yy}),$$

$$\Delta = \frac{\sqrt{2}}{3}\Delta_{\mathrm{so}},$$

with the effective mass parameters A_i, the valence band deformation potentials D_i, which determine the energetic shift due to strain along the crystal directions, and the crystal field and spin-orbit splitting energies Δ_{cr}, Δ_{so}. The deformation potentials of InN are adapted from Ref. [16] (D_1 to D_4) and Ref. [17] ($D_5 = -3.5\,\mathrm{eV}$, $D_6 = -7.1\,\mathrm{eV}$). To implement quantum confinement in a semipolar quantum well, the Hamiltonian has to be transformed into the growth coordinate system (x', y', z') via the rotation matrix U (Eq. 5.2),

$$\begin{pmatrix} k_x \\ k_y \\ k_z \end{pmatrix} \rightarrow \begin{pmatrix} k_{x'} \\ k_{y'} \\ k_{z'} \end{pmatrix} = U \begin{pmatrix} k_x \\ k_y \\ k_z \end{pmatrix} \tag{5.12}$$

As the bulk Hamiltonian has a radial symmetry in the c-plane, the azimuthal angle of the semipolar plane with respect to the m-axis needs not to be taken into account. Within the present model, a- and m-direction of the crystal are equivalent. Different influences of dislocations and differences in strain along those two directions, which may occur in realistic structurs, are not taken into account. Within this approximation, it is sufficient to implement only a polar rotation by the crystal angle θ with the matrix U. The band structure and wave functions of the holes are then obtained by replacing $k_{z'}$ by a differential operator and solving the Schrödinger Eq. 2.3 in the growth coordinate system,

$$\left(\mathbf{H}_h(k_{x'}, k_{y'}, -i\frac{\mathrm{d}}{\mathrm{d}z'}) + V_{VB}(z') \right) \vec{\psi}_f(z') = E_f \vec{\psi}_f(z'). \tag{5.13}$$

Figure 5.7 shows the valence bands of an InGaN quantum well with 35% In-content and 3 nm width on the $(11\bar{2}2)$- and the $(20\bar{2}1)$-plane. The in-plane directions in a semipolar quantum well are not equivalent like in a c-plane QW, so their is no radial symmetry in the quantum well plane and the energy dispersion for holes is different for wave vectors parallel or perpendicular to the c'-direction. The effective hole mass, which is the inverse of the curvature of the band, depends strongly on the direction of the hole wave vector. Therefore, band anti-crossings appear where bands with different effective masses meet. Higher order confinement states in the quantum well give rise to additional bands at lower energy, which also form avoided crossings with the topmost bands.

In contrast to a c-plane QW, the hole eigenfunctions are not heavy hole and light hole states, but they are predominantly polarized along the two directions in the QW plane (compare Sect. 2.3). This is a consequence of the absent radial symmetry and the different strain along the two directions [18]. The topmost bands in a non-c-plane quantum well are therefore not denoted LH- and HH-band, but A- and B-band, depending on their hole wave function at the Γ-point. In the present notation, A denotes the band with a predominant $P_{y'}$-hole polarization and B corresponds to a $P_{x'}$-hole polarization. The band with a predominant hole polarization along the growth direction z' is shifted downwards by several hundred meV with respect to the topmost band due to the compressive strain in the InGaN quantum well layer, similarly to the CH-band in a c-plane QW. Therefore, this polarization does not play a role in GaN-based semipolar optoelectronic devices.

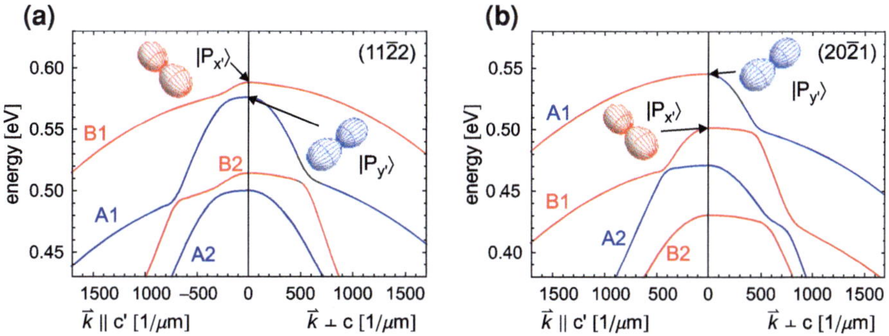

Fig. 5.7 Valence band structure of a 3 nm In$_{0.35}$Ga$_{0.65}$N quantum well on the (11$\bar{2}$2)-plane (**a**) and the (20$\bar{2}$1)-plane (**b**). The angular momentum functions indicate the predominant component of the hole wave function at the Γ-point for the topmost two bands

5.4 Anisotropic Optical Gain

In contrast to a c-plane quantum well, where all in-plane directions are equivalent, the optical gain in a non-c-plane oriented InGaN quantum well is highly anisotropic. This is a consequence of the directional polarization of the hole wave functions in a semi-polar QW and the birefringence of the waveguide material. To calculate the material gain in a semipolar laser diode within the free-carrier-theory, all possible transitions in the band structure are integrated over the quantum well plane, as described in Sect. 2.3:

$$G(\hbar\omega) = \frac{1}{d} \frac{e^2}{4\pi^2 m_0^2 c_0 \varepsilon_0 n_{\text{eff}} \omega E_{\text{hom}}} \sum_{i,f} \int d^2k' \left| M_{if}(\vec{k}') \right|^2 \tag{5.14}$$

$$\cdot \operatorname{sech}\left(\frac{\Delta E_{if}(\vec{k}') - \hbar\omega}{E_{\text{hom}}} \right) \left(f(E_i(\vec{k}'), \mu_e) - f(E_f(\vec{k}'), \mu_h) \right), \tag{5.15}$$

$$M_{if}(\vec{k}') = \langle i \, |\vec{a} \cdot \vec{p}| \, f \rangle (\vec{k}'), \tag{5.16}$$

$$\Delta E_{if}(\vec{k}') = E_i(\vec{k}') - E_f(\vec{k}'). \tag{5.17}$$

The polarization of the hole wave function enters via the matrix element M_{if} and the optical mode polarization vector $\vec{a}$. The matrix element obeys the dipole selection rules:

$$\langle S \, |p_x| \, P_x \rangle = \langle S|p_y|P_y \rangle = \langle S \, |p_z| \, P_z \rangle = \sqrt{\frac{1}{2} E_p m_0}, \tag{5.18}$$

with the momentum matrix element energy E_{p}, the anisotropy of which is neglected here. All other combinations of momentum operator $p_{x,y,z}$ and basis state $\left| P_{x,y,z} \right\rangle$

are equal zero. The absolute squared momentum matrix elements $|\langle i\,|\vec{p}|\,f\rangle\,(\vec{k}')|^2$ are then calculated by multiplying the absolute squared overlap integral of the corresponding wave function components with the momentum matrix element energy E_p times $\frac{1}{2}m_0$. For the choice of basis states made in Eq. 5.10, the momentum matrix elements for spin up states are:

$$\langle i\,|p_x|\,f\rangle = \sqrt{\frac{1}{2}E_p m_0}\left(-\langle\psi_i|\psi_{f,1}\rangle + \langle\psi_i|\psi_{f,2}\rangle\right), \tag{5.19}$$

$$\langle i|p_y|f\rangle = \sqrt{\frac{1}{2}E_p m_0}\left(i\,\langle\psi_i|\psi_{f,1}\rangle + i\,\langle\psi_i|\psi_{f,2}\rangle\right), \tag{5.20}$$

$$\langle i\,|p_z|\,f\rangle = \sqrt{\frac{1}{2}E_p m_0}\,\langle\psi_i|\psi_{f,3}\rangle \tag{5.21}$$

with the component-wise overlap integral of the normalized wave functions:

$$\langle\psi_i|\psi_{f,j}\rangle = \int dz'\,\psi_i^*(z')\psi_{f,j}(z'). \tag{5.22}$$

The same is valid for conduction band spin down states, where the indices $(1, 2, 3)$ have to be replaced by $(4, 5, 6)$. $\psi_{f,j}$ means the j-th component of the hole wave function $\vec{\psi}_f$ of the final state $|f\rangle$, ψ_i is the electron wave function of the initial state $|i\rangle$.

Inhomogeneous broadening due to fluctuations of indium content and QW width is implemented by a constant quasi-Fermi energy model, as described in Ref. [19]. The effect of the fluctuations is treated as a variation in the bandgap energy, neglecting all other consequences on the band structure. The gain is then obtained as an integral over gain curves with different bandgaps weighted with a Gaussian distribution $\Omega(E',\sigma)$ where σ is the inhomogeneous broadening energy:

$$G_{\text{inh}}(\hbar\omega) = \int dE'\,G_{\Delta E\rightarrow\Delta E+E'}(\hbar\omega)\Omega(E',\sigma) \tag{5.23}$$

Here, $G_{\Delta E\rightarrow\Delta E+E'}$ is the material gain calculated from Eq. (5.15), where the band energies E_i and E_f are shifted as follows:

$$E_i(\vec{k}') \rightarrow E_i(\vec{k}') + \eta_c E' \tag{5.24}$$

$$E_f(\vec{k}') \rightarrow E_f(\vec{k}') - \eta_v E' \tag{5.25}$$

with the band offset ratios $\eta_c = 0.74$ and $\eta_v = 0.26$ for the conduction and valence bands, respectively. For the gain calculations presented in this chapter, an inhomogeneous broadening of $\sigma = 50\,\text{meV}$ is assumed.

As described in Sect. 2.3, the free-carrier gain model is suitable for a qualitative comparison of different structures, although it does not allow quantitative predictions for experiments, due to the neglection of many-body effects. To get a more realistic estimate for the emission wavelength, the bandgap renormalization that arises from the many-body effects can be included ad-hoc simply by energetically shifting the calculated gain spectra. The dominant contribution is the Coulomb-hole self energy, which is due to increasingly effective plasma screening at higher carrier density [20]:

$$\Delta E_{\text{CH}} = -2E_R a_0 \int_0^\infty dk' \frac{1}{1 + \frac{k'}{\kappa} + \frac{a_0 k'^3}{32\pi n_{2D}}}. \tag{5.26}$$

A second contribution comes from the Hartree-Fock energy correction, the screened exchange shift:

$$\Delta E_{\text{SX}} = -\sum_{i,f} \frac{E_R a_0}{\pi\kappa} \int d^2k' \frac{1 + \frac{a_0\kappa k'^2}{32\pi n_{2D}}}{1 + \frac{k'}{\kappa} + \frac{a_0 k'^3}{32\pi n_{2D}}}$$
$$\cdot \left(1 - f(E_f(\vec{k}'), \mu_h) + f(E_i(\vec{k}'), \mu_e)\right). \tag{5.27}$$

E_R and a_0 are the exciton Rydberg energy and Bohr radius, respectively and κ is the inverse screening length,

$$a_0 = \frac{4\pi\hbar^2\varepsilon_0\varepsilon}{e^2 m_r}, E_R = \frac{\hbar^2}{2m_r a_0^2}, \tag{5.28}$$

$$\kappa = \frac{e^2}{2\varepsilon\varepsilon_0}\left(\frac{dn_{2D}}{d\mu_e} - \frac{dn_{2D}}{d\mu_h}\right), \tag{5.29}$$

with the static dielectric constant ε, the reduced electron-hole mass $m_r = (m_e^{-1} + m_h^{-1})^{-1} \approx m_e$ (as the effective electron mass is much smaller than the hole mass in InGaN) and the two-dimensional charge carrier density n_{2D} as a function of the quasi-Fermi energies μ_e, μ_h.

Material gain spectra of a 3 nm wide $(20\bar{2}1)$-oriented InGaN QW with 35% indium content are calculated for the two possible waveguide orientations and their optical eigenmodes. On the $(20\bar{2}1)$-plane, the hole wavefunction of the topmost valence band is predominantly polarized along the y' direction (perpendicular to the c-axis). The energetic separation of the topmost band and the second valence band, which has a hole polarization along x' (parallel to the c'-direction), is greater than the thermal energy, so the second band has a significantly lower hole population than the topmost band at room temperature (see Fig. 5.7).

For a waveguide orientation parallel to the c'-direction (see Fig. 5.8a), the electric field vector of the TE-mode points along the y'-direction, which matches the hole polarization of the topmost band, so the gain for this mode is highest. The TM-mode has no gain at relevant charge carrier densities, as the valence band with a z'-polarization is energetically shifted downwards due to the compressive strain of the QW. Because of this energetic shift, the quantum well is nearly transparent

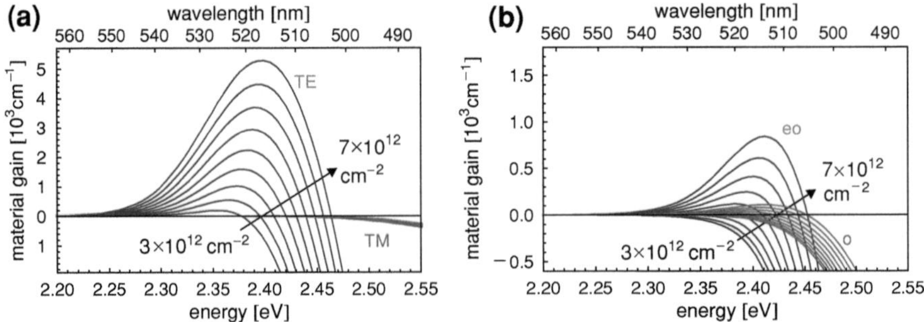

Fig. 5.8 Material gain spectra for corresponding optical mode polarizations of a 3 nm $In_{0.35}Ga_{0.65}N$ QW on the $(20\bar{2}1)$-plane for a waveguide orientation parallel to c' (**a**) and perpendicular to the c-direction (**b**) for charge carrier densities of 3–7×10^{12}cm^{-2} in steps of 0.5×10^{12}cm^{-2}. Different vertical scales are plotted for (**a**) and (**b**)

for TM-polarized light at the laser wavelength (which is the peak wavelength of the TE-gain). Absorption of TM-polarized light sets in at higher energies, where transitions to the z'-polarized valence band are available.

The eo-mode (see Fig. 5.8b), which occurs for a waveguide orientation perpendicular to the c-axis, has much less gain than the TE-mode as transitions to the topmost band do not contribute to the optical polarization along the extraordinary direction. The corresponding valence band for the eo-mode is only the second band, which generally has less population than the topmost band due to the Fermi-Dirac distribution of the holes among the individual bands. Correspondingly, the gain peak for the eo-mode is shifted to blue by several ten meV. An additional reduction of the gain on the extraordinary mode comes from the mismatch of hole polarization, which is in-plane due the compressive strain, and optical polarization, which points along the c-direction. For the $(20\bar{2}1)$-plane, this causes a reduction of the gain by a factor $\cos^2(15°) \approx 0.93$. The ordinary mode has a high angular mismatch to both hole polarizations of the topmost valence bands, so it has negligible gain.

For comparison, material gain spectra of a c-plane oriented quantum well with 28% indium content are shown in Fig. 5.9. The high internal fields cause a red-shift of the gain peak by several hundred meV due to the quantum-confined Stark effect. Therefore, a significantly lower indium content is required to achieve emission in the green spectral range for c-plane quantum wells than for semipolar crystal orientations [21]. The overall gain is very low compared to the semipolar quantum well, owing to the drastic reduction of the wavefunction overlap in c-plane quantum wells. There is also a pronounced blue-shift of the gain peak with increasing excitation, which relates to the partial screening of the internal fields by charge carriers.

The anisotropic optical gain causes a dependence of the laser characteristics on the waveguide orientation in semipolar devices. Laser diodes with a waveguide perpendicular to the c-axis have a higher threshold and shorter emission wavelength than devices with a waveguide along c'. This finding has been confirmed experimentally

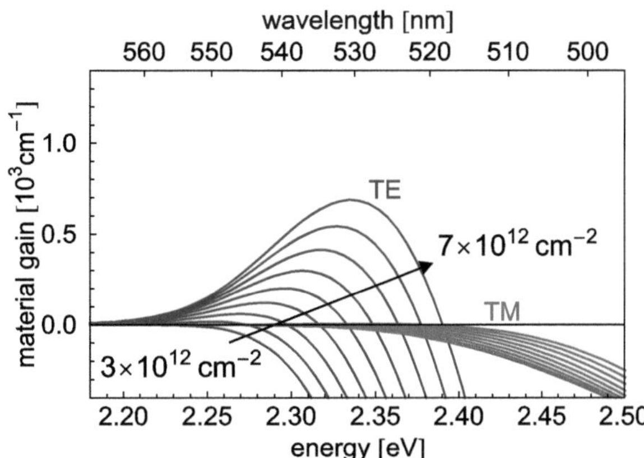

Fig. 5.9 Material gain spectra for TE- and TM-mode of a c-plane oriented 3 nm $In_{0.28}Ga_{0.72}N$ QW for charge carrier densities of 3–7 $\times$ 10^{12}cm^{-2} in steps of 0.5 $\times$ 10^{12}cm^{-2}

in Ref. [22], where the threshold current of semipolar green laser diodes differed by a factor of four for the two possible orientations.

5.5 Polarization Switching

For quantum wells on the $(20\bar{2}1)$-plane, the separation of the topmost bands is several ten meV and the hole wave function of the topmost band is oriented along y', so the dominant optical polarization is always perpendicular to the c-axis. On the other hand, for $(11\bar{2}2)$-oriented quantum wells, the separation of the topmost two bands is small and an avoided crossing occurs very close to the Γ-point (see Fig. 5.7). The relative positions of the individual valence bands are determined by the strain shifts. On the $(11\bar{2}2)$-plane, the dominant strain tensor component is the shear strain ϵ_{xz} (see Fig. 5.3), which energetically lifts the B-band and lowers the A-band [23]. If the energetic shift induced by the shear strain is sufficient, then the order of the bands is changed, and consequently, the topmost band changes its hole polarization, as illustrated in Fig. 5.10. As the strain increases with increasing indium content, this means that the hole polarization of the topmost band, and thereby the dominant optical polarization of the emission, is a function of the indium content in the QW. For the choice of band parameters employed here, a switching of the hole polarization occurs at an indium content of about 28%, which agrees well with experimental data on polarization switching in semipolar LEDs [23]. This effect does not occur on the $(20\bar{2}1)$-plane, as the shear strain is much lower on this plane. It is pointed out in Ref. [23], that in order for the theory to reproduce the experimentally measured polarization switching, a shear strain deformation potential D_6 for InN has to be used which is significantly higher than the one for GaN.

The switching of the hole polarization of the topmost valence band in $(11\bar{2}2)$-oriented QWs has a considerable impact on the optical gain of laser diodes grown on this plane. However, in order to calculate the optical gain, one also has to take into

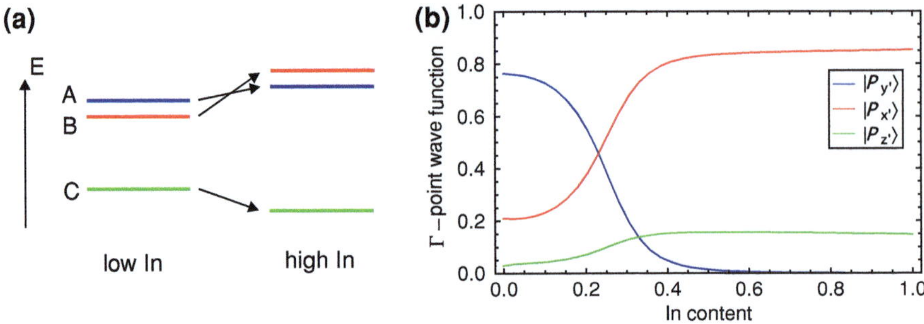

Fig. 5.10 Illustration of band order in a $(11\bar{2}2)$-oriented quantum well at low and high indium content (**a**) and wave function components of the topmost band as function of indium content (**b**)

account the wave vector-dependence of the hole wave functions, as higher energy states also contribute to the optical gain at charge carrier densities relevant for laser diodes. In semipolar quantum wells, band anti-crossings occur close to the Γ-point due to the missing rotational symmetry of the effective hole mass. Consequently, a band does not retain its wave function composition all over k-space. The wave function compositions and thus also the interband momentum matrix elements of the bands then depend on the in-plane wave vector of the charge carriers.

Figure 5.11 shows band structures and hole polarizations of an InGaN quantum well on the $(11\bar{2}2)$-plane as functions of $k_{x'}$ and $k_{y'}$ for three different indium contents. The simulated quantum well thickness is 3 nm. Here, the hole polarization is indicated by the difference between the interband momentum matrix elements for TE-polarization, M_{TE}^2, and extraordinary polarization, M_{eo}^2. The anti-crossing close to the Γ-point, which occurs at small values of $k_{y'}$ at low indium content (see Fig. 5.11a) shifts through the Γ-point and to low values of $k_{x'}$ with increasing indium content due to the increasing shear strain. Consequently, the Γ-point polarization switches, as discussed above. However, a closer look at the polarization graphs reveals that only a small part of the band changes its wave function, so higher states of the band still contribute to emission on the other direction. Furthermore, the matrix element M_{eo}^2 does not reach the same magnitude as M_{TE}^2, because the hole wave functions are nearly aligned along the x'-direction in the quantum well and not along the direction of the optical polarization of the extraordinary mode. The matrix element is thus reduced by a cosine square of the angle between x'-direction and c-axis ($32°$ for $(11\bar{2}2)$-plane).

Figure 5.12 shows optical gain spectra for TE- and extraordinary modes for an InGaN/GaN quantum well on the $(11\bar{2}2)$-plane with low and high indium content. At an indium content lower than 28%, the TE-mode has significantly higher gain then the extraordinary mode, whereas for higher indium content, the gain on the eo-mode reaches about the same magnitude as the TE-gain. However, the eo-gain does not exceed the TE-gain at elevated charge carrier densities, although the valence subband which contributes to eo-emission at the Γ-point is in this case the topmost

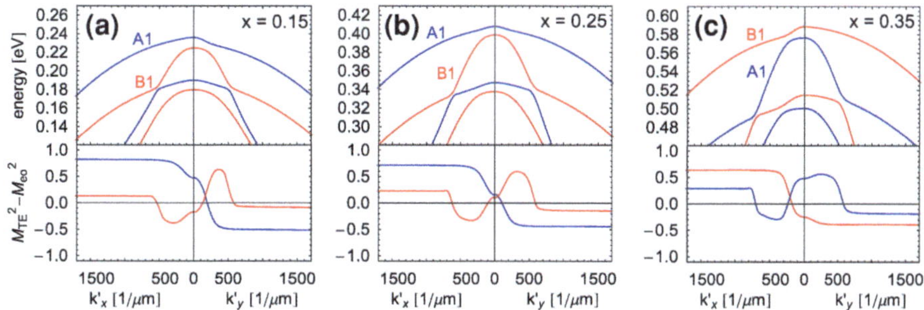

Fig. 5.11 Hole energies and polarizations $M_{\mathrm{TE}}^2 - M_{\mathrm{eo}}^2$ of the two topmost valence bands A1 (*blue*) and B1 (*red*) versus wavenumber along two orthogonal in-plane directions for 15% (**a**), 25% (**b**) and 35% (**c**) indium content in a ($11\bar{2}2$)-oriented QW. The polarization $M_{\mathrm{TE}}^2 - M_{\mathrm{eo}}^2$ is given in units of the momentum matrix element parameter $\frac{1}{2}E_{\mathrm{p}}m_0$. Higher order valence bands are drawn for reasons of clarity

Fig. 5.12 Optical gain spectra for TE-mode (*top*) and extraordinary mode (*bottom*) of a 3 nm InGaN quantum well with 15% (*left*) and 35% indium content (*right*) on the semipolar ($11\bar{2}2$)-plane. Sheet charge carrier densities are 3.0–$6.5 \times 10^{12} \mathrm{cm}^{-2}$ (from *bottom* to *top curves*) in steps of $0.5 \times 10^{12} \mathrm{cm}^{-2}$

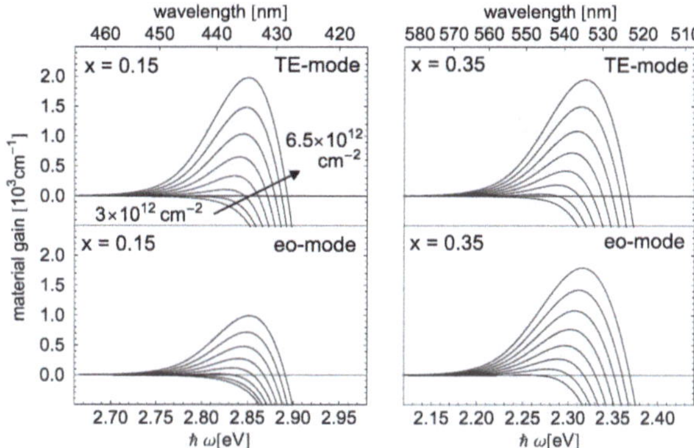

band (compare Fig. 5.11). In this case, the highest gain arises then from transitions to the second topmost valence band, which has less hole population than the topmost one, but a significantly higher transition matrix element.

Experimental investigations of the anisotropic optical gain in laser diodes grown on the ($11\bar{2}2$)-plane emitting at 500 nm showed that the TE-mode still has higher gain at the corresponding indium content, even though the dominant polarization of the spontaneous emission is switched to the c'-direction [24]. This is a consequence of the misalignment of photon polarization and valence band hole wavefunction for eo-modes which reduces the transition matrix element, as discussed above. The transparency threshold for the eo-mode is thus somewhat lower than for the TE-mode, but the differential gain, that is the increase in gain per charge carrier density, is less. This effect was measured in Ref. [24], where it was attributed to spatial fluctuations of the indium composition of the quantum wells. The calculations presented here show that this phenomenon is inherent to the band structure of ($11\bar{2}2$)-oriented InGaN QWs with high indium content. However, spatial inhomogeneities, which are only

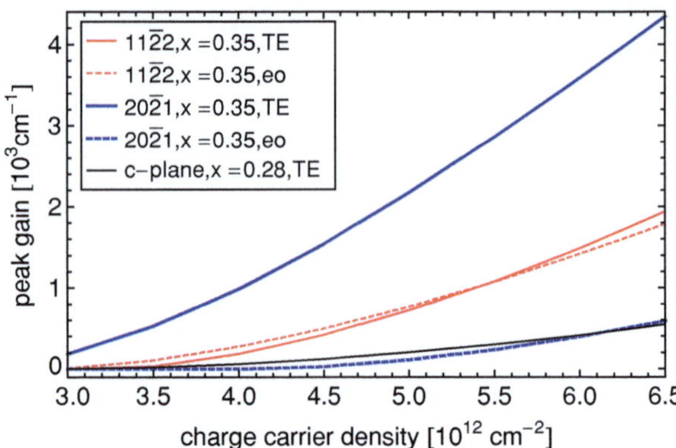

Fig. 5.13 Peak gain as function of charge carrier density for TE- and eo-modes in a semipolar 3 nm $In_{0.35}Ga_{0.65}N$ quantum well on the $(11\bar{2}2)$- and the $(20\bar{2}1)$-plane, and for the TE-mode in a 3 nm $In_{0.28}Ga_{0.72}N$ QW on the c-plane

treated as fluctuations in the band gap in this calculation, further enhance this effect [25].

At an indium content of 35%, which is suitable for true green emission around 530 nm, semipolar laser diodes on the $(11\bar{2}2)$-plane have approximately the same optical gain for both possible waveguide orientations. Orienting the laser waveguide perpendicular to the c-axis, which offers a low-index plane for mirror cleavage, would thus not significantly decrease the optical gain of such devices.

5.6 Comparison of Crystal Planes

Owing to the reduction of the internal fields, semipolar InGaN quantum wells with an indium content suitable for green emission exhibit considerably higher optical gain at a given charge carrier density than corresponding c-plane quantum wells. Figure 5.13 shows the peak gain as a function of charge carrier density for different semipolar planes and their possible waveguide orientations compared to a c-plane QW. An inhomogeneous broadening of $\sigma = 50$ meV is assumed for all orientations. In the $(20\bar{2}1)$-oriented QW, positive gain sets in already at a very low carrier density due to the large energetic separation of the topmost valence bands. At an elevated charge carrier density, the TE-mode gain is by a factor of six higher than in a c-plane QW with comparable emission wavelength. However, the improvement of the optical gain only applies to the waveguide orientation along the c'-direction, the fundamental optical mode of which has a TE-polarization. The eo-mode of the other waveguide orientation has much less gain, comparable to the c-plane QW. On the other hand, the optical gain in a quantum well on the $(11\bar{2}2)$-plane is lower than on the $(20\bar{2}1)$-plane, but still by a factor of three higher than in the c-plane quantum well. Owing to the polarization switching, TE-mode and eo-mode have similar gain, so both waveguide orientations can be used.

It has to be noted that in general, the improvements in optical gain in laser diodes on the semipolar planes will not proportionally decrease the threshold current, as the improved wave function overlap also leads to an increase in the spontaneous emission, and possibly also the Auger recombination, which results in a substantial decrease of the charge carrier lifetime. Nevertheless, vast improvements can be expected for semipolar green laser diodes, as soon as similar material quality and heterostructure design as in state-of-the-art c-plane devices are achieved.

References

1. T. Miyoshi, S. Masui, T. Okada, T. Yanamoto, T. Kozaki, S.-I. Nagahama, T. Mukai, 510–515 nm InGaN-based green laser diodes on c-plane GaN substrate. Appl. Phys. Express **2**, 062201 (2009)
2. A. Avramescu, T. Lermer, J. Müller, C. Eichler, G. Bruederl, M. Sabathil, S. Lutgen, U. Strauss, True green laser diodes at 524 nm with 50 mW continuous wave output power on c-plane GaN. Appl. Phys. Express **3**, 061003 (2010)
3. J. W. Raring, M. C. Schmidt, C. Poblenz, B. Li, Y.-C. Chang, M. J. Mondry, Y.-D. Lin, M. R. Krames, R. Craig, J. S. Speck, S. P. DenBaars, S. Nakamura, High-performance blue and green laser diodes based on nonpolar/semipolar bulk GaN substrates. Proc. SPIE **7939**, 79390Y (2011)
4. M. Adachi, Y. Yoshizumi, Y. Enya, T. Kyono, T. Sumitomo, S. Tokuyama, S. Takagi, K. Sumiyoshi, N. Saga, T. Ikegami, M. Ueno, K. Katayama, T. Nakamura, Low threshold current density InGaN based 520–530 nm green laser diodes on semi-polar $(20\bar{2}1)$ free-standing GaN substrates. Appl. Phys. Express **3**(12), 121001 (2010)
5. T. Akiyama, T. Yamashita, K. Nakamura, T. Ito, Stability and indium incorporation processes on $In_{0.25}Ga_{0.75}N$ surfaces under growth conditions: first-principles calculations. Jpn. J. Appl. Phys. **49**, 030212 (2010)
6. J. E. Northrup, GaN and InGaN$(11\bar{2}2)$ surfaces: group-III adlayers and indium incorporation. Appl. Phys. Lett. **95**,133107 (2009)
7. I. Vurgaftman, J. Meyer, Electron bandstructure parameters. In: Piprek J. (ed.), Nitride semiconductor devices: principles and simulations, chapter 2, Wiley VCH, Weinheim p. 13–48 (2007)
8. S. Park, S. Chuang, Crystal-orientation effects on the piezoelectric field and electronic properties of strained wurtzite semiconductors. Phys. Rev. B **59**(7), 4725–4737 (1999)
9. O. Ambacher, J. Majewski, C. Miskys, A. Link, M. Hermann, M. Eickhoff, M. Stutzmann, F. Bernardini, V. Fiorentini, V. Tilak, B. Schaff, L. F. Eastman, Pyroelectric properties of Al(In)GaN/GaN hetero- and quantum well structures. J. Phys.: Condens. Matter **14**, 3399–3434 (2002)
10. W. Scheibenzuber, U. Schwarz, R. Veprek, B. Witzigmann, A. Hangleiter, Calculation of optical eigenmodes and gain in semipolar and nonpolar InGaN/GaN laser diodes. Phys. Rev. B **80**(11), 115320 (2009)
11. S. Shokhovets, R. Goldhahn, G. Gobsch, S. Piekh, R. Lantier, A. Rizzi, V. Lebedev, W. Richter, Determination of the anisotropic dielectric function for wurtzite AlN and GaN by spectroscopic ellipsometry. J Appl. Phys. **94**(1), 307 (2003)
12. M. Vasell, Structure of guided modes in planar multilayers of optically anisotropic materials. J. Opt. Soc. Am. **64**, 166 (1974)
13. J. Rass, T. Wernicke, W. G. Scheibenzuber, U. T. Schwarz, J. Kupec, B. Witzigmann, P. Vogt, S. Einfeldt, M. Weyers, M. Kneissl, Polarization of eigenmodes in laser diode waveguides on semipolar and nonpolar GaN. Physica Status Solidi RRL **4**(1), 1–3 (2010)

14. C.-Y. Huang, A. Tyagi, Y.-D. Lin, M. T. Hardy, P. S. Hsu, K. Fujito, J.-S. Ha, H. Ohta, J. S. Speck, S. P. DenBaars, S. Nakamura, Propagation of spontaneous emission in birefringent m-axis oriented semipolar $(11\bar{2}2)$ (Al,In,Ga)N waveguide structures. Jpn. J. Appl. Phys. **49**,010207 (2010)
15. S. Chuang, C. Chang, $k \cdot p$ method for strained wurtzite semiconductors. Phys. Rev. B **54**(4),2491–2504 (1996)
16. B. Gil, M. Moret, O. Briot, S. Ruffenach, C. Giesen, M. Heuken, S. Rushworth, T. Leese, M. Succi, InN excitonic deformation potentials determined experimentally. J. Cryst. Growth **311**, 2798–2801 (2009)
17. M. Funato, Y. Kawakami, Polarization anisotropy in semipolar/polar nitride semiconductor quantum wells. invited talk at ICNS-8, South Korea (2009)
18. K. Kojima, H. Kamon, M. Funato, Y. Kawakami, Optical anisotropy control of non-c-plane InGaN quantum wells. Jpn. J. Appl. Phys. **48**(8), 080201 (2009)
19. K. Kojima, U. T. Schwarz, M. Funato, Y. Kawakami, S. Nagahama, T. Mukai, Optical gain spectra for near UV to aquamarine (Al, In)GaN laser diodes. Opt. Express **15**, 7730–7736 (2007)
20. W. W. Chow, S. W. Koch, *Semiconductor-Laser Fundamentals* (Springer, Berlin, 1998)
21. A. Avramescu, T. Lermer, J. Müller, S. Tautz, D. Queren, S. Lutgen, U. Strauss, InGaN laser diodes with 50 mW output power emitting at 515 nm. Appl. Phys. Lett. **95**, 071103 (2009)
22. T. Kyono, Y. Yoshizumi, Y. Enya, M. Adachi, S. Tokuyama, M. Ueno, K. Katayama, T. Nakamura, Optical polarization characteristics of InGaN quantum wells for green laser diodes on semi-polar $(20\bar{2}1)$GaN substrates. Appl. Phys. Express **3**,011003 (2010)
23. M. Ueda, M. Funato, K. Kojima, Y. Kawakami, Y. Narukawa, T. Mukai, Polarization switching phenomena in semipolar $In_x Ga_{1-x}$ N/GaN quantum well active layers. Phys. Rev. B **78**,233303 (2008)
24. D. S. Sizov, R. Bhat, J. Napierala, C. Gallinat, K. Song, C.-E. Zah, 500-nm optical gain anisotropy of semipolar $(11\bar{2}2)$ InGaN quantum wells. Appl. Phys. Express **2**,071001 (2009)
25. K. Kojima, A. A. Yamaguchi, M. Funato, Y. Kawakami, S. Noda, Gain Anisotropy Analysis in Green Semipolar InGaN Quantum Wells with Inhomogeneous Broadening. Jpn. J. Appl. Phys. **49**(8), 081001 (2010)

Chapter 6
Dynamics of Charge Carriers and Photons

In order to achieve a low threshold in GaN-based laser diodes, sufficient optical gain has to be provided by the active region to compensate the optical losses at a low pump current. The optical gain is a function of charge carrier density in the quantum wells, which is proportional to the product of current and charge carrier lifetime. A reduction of threshold current can thus be achieved by optimizing the active region for a long carrier lifetime.

This chapter treats the dependency of the charge carrier lifetime on the carrier density in the quantum wells and presents a method to determine the individual recombination coefficients from the dynamical properties of a laser diode. In particular, a clear differentiation is made between charge carrier leakage and recombination. This topic also has high relevance for GaN-based light emitting diodes (LEDs), where higher-order non-radiative recombination and carrier leakage limit the efficiency already at moderate currents [1–3]. Additionally, gain saturation and differential gain are deduced from relaxation oscillations using the rate equation model. The obtained recombination coefficients are used to analyze the dependency of optical gain on charge carrier density experimentally and verify the utilized linear gain model.

The sample studied in this chapter is a violet laser diode emitting at around 415 nm, grown on c-plane oriented free-standing GaN substrate at the Ecole Polytechnique Fédérale de Lausanne (EPFL) by metal-organic vapor phase epitaxy. It has a ridge width of 3 μm, a cavity length of 600 μm and uncoated facets. The active region consists of a p-AlGaN electron blocking layer and two 4.5 nm wide InGaN quantum wells separated by 12 nm undoped InGaN barriers with low In-content. AlGaN cladding layers are used for transversal waveguiding. The confinement factor

W. G. Scheibenzuber, *GaN-Based Laser Diodes*, Springer Theses,
DOI: 10.1007/978-3-642-24538-1_6, © Springer-Verlag Berlin Heidelberg 2012

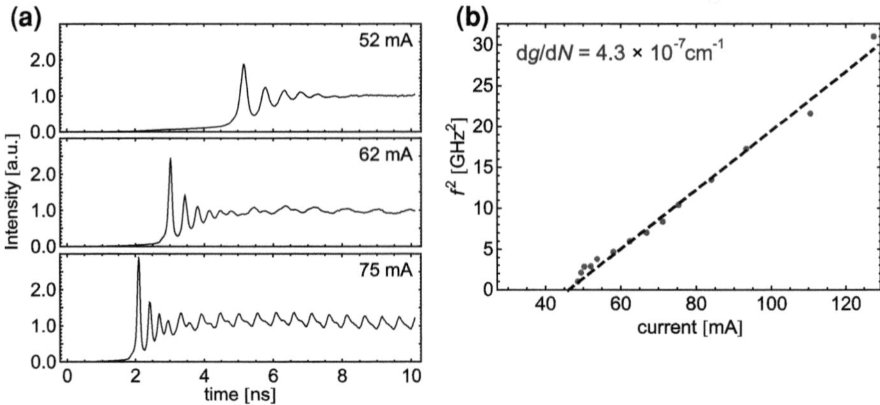

Fig. 6.1 **a** Laser emission intensity as a function of time for pump currents of 52 mA (*top*), 62 mA (*center*) and 75 mA (*bottom*). The electrical excitation pulse is turned on at $t = 0$. **b** Relaxation oscillation frequency squared versus pump current (data points) and linear fit to the data (*dashed line*)

of the optical mode is $\Gamma = 0.029$, calculated with a two-dimensional waveguide solver.

6.1 Differential Gain and Gain Saturation

The relaxation behavior of a laser diode upon turn-on is determined by several internal parameters, such as the differential gain, the gain saturation and the charge carrier lifetime. These parameters can be obtained from a systematic analysis of the frequency and damping of the relaxation oscillations and the turn-on delay (see Sect. 2.4). These properties are measured using a streak camera with a temporal resolution better than one percent of the set time range. Solving Eqs. 2.15 and 2.16 for small perturbations ΔN and ΔS from steady state yields the functional dependence of the oscillation frequency on pump current [4],

$$f_r = \frac{1}{2\pi} \sqrt{\frac{dg}{dN} \eta_{\text{inj}} \frac{c}{e} (I - I_{\text{th}})}, \tag{6.1}$$

The differential gain per carrier number, dg/dN is defined as the derivative of the modal gain with respect to the number of charge carriers in the active region, evaluated at the laser threshold and the laser emission wavelength. It cannot be obtained from optical gain spectroscopy alone, because that method does not allow an estimation of the charge carrier number. As can be seen from Eq. 6.1, dg/dN can be determined from a linear fit of the square of the oscillation frequency to the pump current. Figure 6.1 illustrates this procedure.

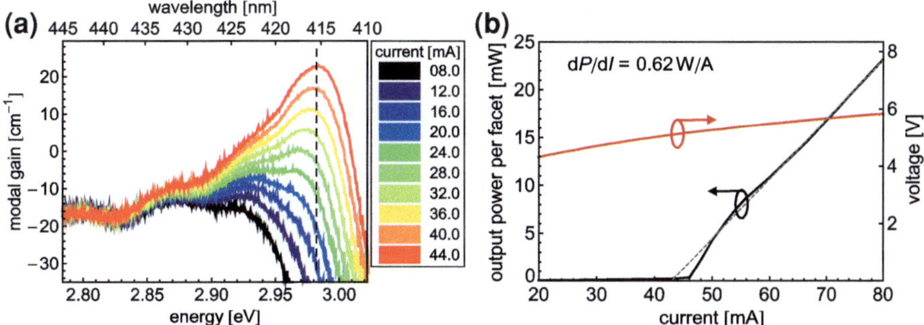

Fig. 6.2 **a** Optical gain spectra at currents from 8 to 44 mA in steps of 4 mA (*from bottom to top*). The dashed line marks the laser wavelength 415.5 nm. **b** Optical output power per facet and operating voltage in continuous wave operation. The dashed gray line is a linear fit to the output power above threshold

Table 6.1 Device parameters extracted from high resolution electroluminescence spectroscopy and output power characteristics		
α_{m} [cm^{-1}]		30 ± 1
α_{int} [cm^{-1}]		18 ± 1
dg/dI [cm^{-1} mA^{-1}]		1.5 ± 0.1
η_{inj}		0.68 ± 0.02
n_{gr}		3.84 ± 0.05

As Eq. 6.1 contains the speed of light in the medium c and the injection efficiency η_{inj}, these parameters have to be measured by high resolution electroluminescence spectroscopy (see Sects. 4.2 and 4.3) in order to extract the differential gain form the relaxation frequency. Figure 6.2 shows optical gain spectra and output characteristics, from which the injection efficiency is extracted. For this sample, the internal losses are $\alpha_{\mathrm{int}} = 18 \pm 1\,\mathrm{cm}^{-1}$ and the mirror losses of the two uncoated facets, the reflectivity of which is approximately 17%, are $\alpha_{\mathrm{m}} = 30 \pm 1\,\mathrm{cm}^{-1}$. The slope efficiency is $0.62\,\mathrm{W/A}$ per facet, which corresponds to an injection efficiency of $(68 \pm 2)\%$. This value indicates an insufficient function of the electron blocking layer and a considerable loss of charge carriers due to electron overflow. The group refractive index is calculated as $n_{\mathrm{gr}} = 3.84 \pm 0.05$ from the spacing of adjacent longitudinal modes, which is 37 pm at the laser wavelength. Table 6.1 summarizes the extracted parameters. Inserting these parameters into Eq. 6.1 yields a differential gain per carrier number of $dg/dN = 4.3 \times 10^{-7}\,\mathrm{cm}^{-1}$, which corresponds to $d\tilde{g}/dN = 7.0 \times 10^{-18}\,\mathrm{cm}^{-1}/\mathrm{cm}^{-3}$ in the charge carrier density dependent notation. The differential gain per current at the laser wavelength 415.5 nm, extracted from optical gain spectroscopy, is $dg/dI|_{\lambda=\mathrm{const.}} = 1.5 \pm 0.1\,\mathrm{cm}^{-1}\,\mathrm{mA}^{-1}$.

The gain saturation parameter k_{sat} describes the reduction of the gain at high photon density (compare Eq. 2.18) due to spectral hole burning [4]. It acts as a damping term for the relaxation oscillations, and can thus be determined from a comparison of the time-dependent solution of the rate equations to the relaxation

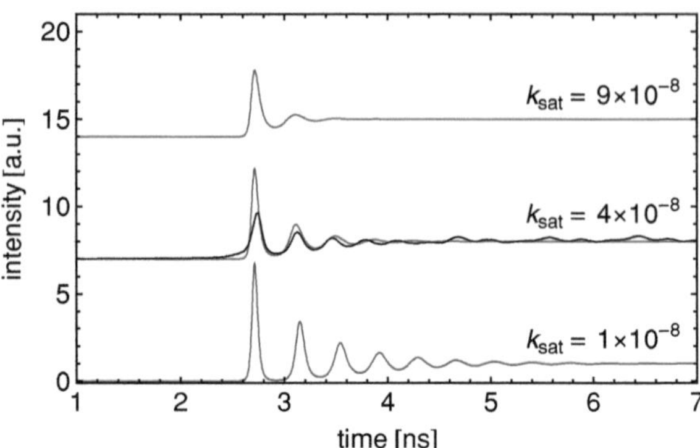

Fig. 6.3 Simulated time evolution of photon number in the cavity for $k_{sat} = 1, 4, 9 \times 10^{-8}$ (*bottom* to *top*) at a pump current of 67 mA. Curves are shifted vertically for clarity. The black line shows the experimental intensity trace at 67 mA

oscillations. Figure 6.3 shows time evolutions of the photon number in the cavity, simulated using Eqs. 2.15 and 2.16, for different values of k_{sat} compared to the experiment. The best agreement is achieved for $k_{sat} = (4 \pm 1) \times 10^{-8}$. To obtain a parameter that is independent of device geometry, this value has to be transformed into a photon-density dependent quantity. Therefore, it is multiplied by the active region volume and divided by the confinement factor, yielding a value of $\tilde{k}_{sat} = (2.2 \pm 0.6) \times 10^{-17}\,\text{cm}^3$.

6.2 Charge Carrier Recombination

GaN-based light emitters suffer from a reduction of their internal quantum efficiency with increasing current, which becomes more pronounced with increasing emission wavelength. It has been debated whether this reduction of efficiency is to be related either to charge carrier leakage [5, 6] or to Auger recombination [7–9]. However, many of the reported results are based only on fitting efficiency curves, which apparently works well with several different models [10]. To clarify this issue, the dependency of the charge carrier lifetime on current has to be analyzed. It is also promising to study recombination in laser diodes [11, 12] instead of LEDs because the dynamical behavior above laser threshold is strongly influenced by the carrier recombination rate. Moreover, laser diodes have the advantage of allowing a direct measurement of their injection efficiency via optical gain spectroscopy. This offers the possibility to clearly differentiate between charge carrier leakage and intrinsic recombination mechanisms.

6.2.1 Charge Carrier Lifetime at Low Excitation

Within the ABC-model (see Sect. 2.4), the functional dependency of the inverse of the charge carrier lifetime $\tau(N)$, which is denoted $r(N)$, on the carrier number N is

$$r(N) := \frac{R(N)}{N} = \frac{1}{\tau(N)} = A + BN + CN^2. \tag{6.2}$$

Here, $R(N)$ denotes the total number of carriers recombining per unit time and $r(N)$ is the recombination rate. In the limit of low excitation ($N \to 0$), the charge carrier lifetime is constant. In this regime, the recombination of charge carriers is approximately exponential with a decay time of A^{-1}, the inverse of the Schockley-Read-Hall parameter. The parameter A scales with the density of threading dislocations in the active region [13]. It can be determined from the decay of the spontaneous emission at low current [12]. An electric pulse generator with a rise/fall time < 100 ps and a streak camera are used to measure the decay of the electroluminescence, which relates to the decay of the charge carriers in the active region via the spontaneous emission rate BN^2. This is only valid at low excitation, as long as stimulated emission is negligible.

Figure 6.4 shows the decaying spontaneous emission after a 50 ns current pulse at 2 mA. Initially, the charge carrier density is still high enough for the BN^2-term to contribute, so the decay is not exponential. But around 20 ns after the trailing edge of the electric pulse the curve approaches a linear behavior on a semi-logarithmic scale, which means that at this time, the charge carrier number in the active region is sufficiently low for the defect-related A-term to be dominant. A linear fit of the curve in this time range gives the electroluminescence decay time τ_{EL}. The defect-related charge carrier lifetime is given by

$$\tau_{\mathrm{SRH}} = 1/A = 2\tau_{\mathrm{EL}}. \tag{6.3}$$

The factor of two arises from the N^2 dependency of the spontaneous emission. Here, $\tau_{\mathrm{SRH}} = 24 \pm 2$ ns is obtained, which corresponds to $A = (4.2 \pm 0.4) \times 10^7 s^{-1}$.

6.2.2 Determination of the Recombination Rate

Assuming the validity of the ABC-model, the recombination rate $r(N)$ (Eq. 6.2) is a parabola, which is generally fixed by three independent constraints. By deriving such constraints for $r(N)$, the recombination coefficients A, B and C can thus be determined. $A = r(0)$ is derived from the decay of the electroluminescence at low excitation, so two more constraints are needed. As differential gain per carrier number $\mathrm{d}g/\mathrm{d}N$ and per current $\mathrm{d}g/\mathrm{d}I$ are related to each other via the charge carrier lifetime, these two quantities can be combined to a constraint for the recombination rate in

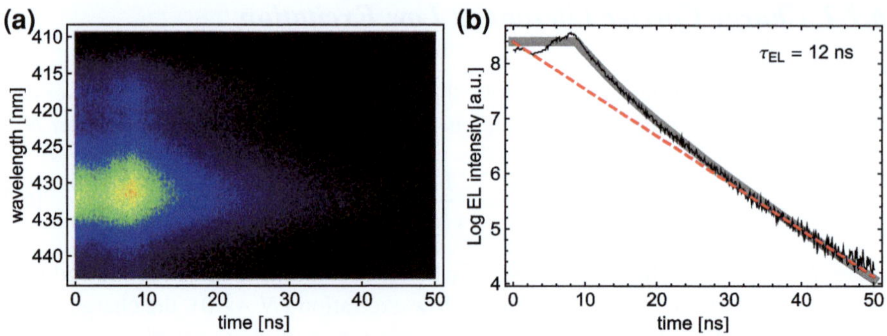

Fig. 6.4 Streak camera trace of the electroluminescence decay after a 2 mA pulse which is switched off at $t = 8.8$ ns (**a**) and spectrally integrated signal on semilogarithmic scale (**b**, *black line*). The dashed red line is a linear fit to the measurement in the time range from 30 to 45 ns, which gives the electroluminescence decay time τ_{EL}. The gray line is the result of a simulation using the A, B and C parameters determined in this work. Deviations from a constant intensity in the time range from 0 to 8 ns are due to the imperfect shape of the electrical excitation pulse

the following way: In a first step, the derivative of the carrier number by the current dN/dI is calculated by

$$\frac{dg}{dI} = \frac{dg}{dN}\frac{dN}{dI}.\tag{6.4}$$

The pump current in steady state has to compensate the carrier recombination:

$$I = \frac{e}{\eta_{\text{inj}}}\frac{N}{\tau} = \frac{e}{\eta_{\text{inj}}}Nr.\tag{6.5}$$

Recombination due to stimulated emission is neglected here, as below threshold it is at least two orders of magnitude smaller than N/τ. This relates dN/dI to the recombination rate:

$$\frac{dI}{dN} = \frac{e}{\eta_{\text{inj}}}\left(r + N\frac{dr}{dN}\right).\tag{6.6}$$

Combining Eqs. 6.4 and 6.6 yields a relation for the slope of Eq. 6.2:

$$\frac{dr}{dN}(N_{\text{th}}) = \left(\eta_{\text{inj}}\frac{dg}{dN}\Big/\left(e\frac{dg}{dI}\right) - \frac{1}{\tau(N_{\text{th}})}\right)\Big/N_{\text{th}}.\tag{6.7}$$

This relation is evaluated at threshold, with the threshold carrier number

$$N_{\text{th}} = \tau(N_{\text{th}})\frac{\eta_{\text{inj}}I_{\text{th}}}{e}.\tag{6.8}$$

The benefit of this step is that it reduces the problem to finding the charge carrier lifetime at threshold $\tau_{\text{th}} := \tau(N_{\text{th}})$. The values of the parameters dg/dI, dg/dN and η_{inj}

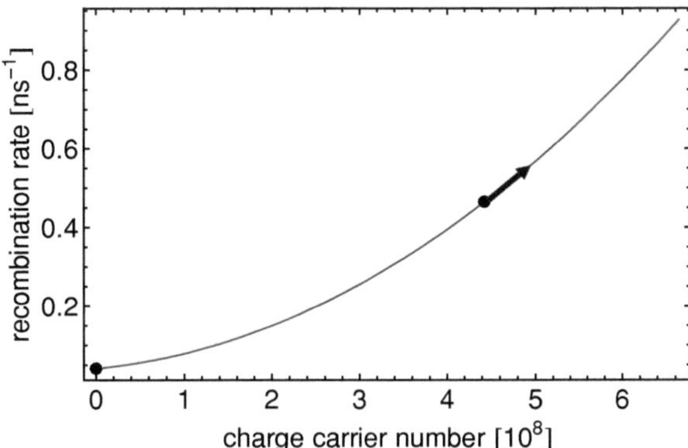

Fig. 6.5 Recombination rate as a (*parabolic*) function of charge carrier number within the ABC model. The black points and arrow indicate the three constrains which are used to fix the parabola

at threshold or slightly below are determined experimentally, as discussed in Sect. 6.1. If τ_{th} is known then the threshold carrier number is fixed by Eq. 6.8 and the slope of $r(N)$ at N_{th} is expressed in terms of known quantities by Eq. 6.7. Then the three constraints $r(0) = A$, $r(N_{th}) = 1/\tau_{th}$ and $dr/dN(N_{th})$ fix the parabola $r(N)$ and thus also the remaining recombination coefficients B and C. Figure 6.5 illustrates this procedure.

6.2.3 Charge Carrier Lifetime at Threshold

The missing quantity τ_{th} can be obtained by fitting the current dependency of the turn-on delay. Therefore, the parameters A, B and C are calculated for different values of τ_{th}. The range of possible values is limited by the requirement that B and C have to be positive. For the present sample, the minimum possible value of τ_{th} is 1.59 ns, which corresponds to $C = 0$, and the maximum value is 2.29 ns, which corresponds to $B = 0$.

The resulting values for A, B and C are inserted into the rate equation model (Eqs. 2.15–2.17) and the model is solved numerically for the current range measured in the experiment. The turn-on delays of the simulated time evolutions are then compared to the experimental ones and from this comparison, the value for τ_{th} is selected which provides the best accordance of simulation and experiment. Figure 6.6 illustrates this procedure. Here, a lifetime at threshold of 2.15 ns gives the best fit to the experimental data. The A, B and C coefficients calculated from this value also reproduce the electroluminescence decay data (compare Fig. 6.4), which proves the consistency of the method.

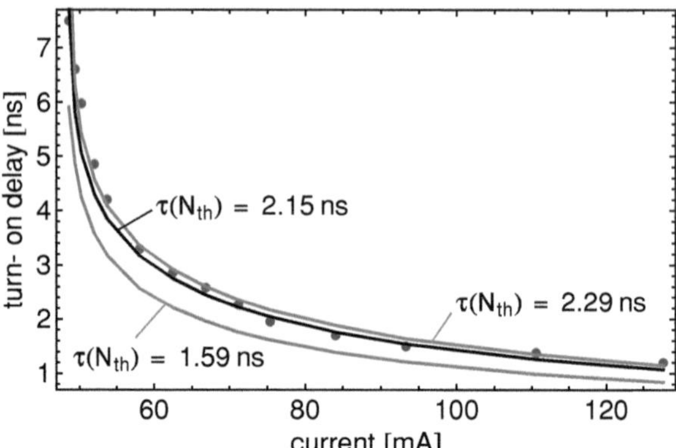

Fig. 6.6 Measured (*points*) and simulated (*lines*) turn-on delay as a function of pump current. The black line is the best fit to the data points. It is calculated with a lifetime at threshold of $\tau_{th} = 2.15$ ns. The gray lines are calculated for the lowest (*lower curve*) and highest (*upper curve*) possible value of τ_{th}

Table 6.2 SRH-coefficient A, spontaneous emission coefficient B, Auger coefficient C, differential gain $\mathrm{d}g/\mathrm{d}N$, gain saturation parameter k_{sat}, transparency carrier number/density N_{tr}, threshold carrier number/density N_{th} and charge carrier lifetime at threshold τ_{th} in volume density (3D), area density (2D) and number notation with relative error for each quantity

Parameter	Number notation	2D notation	3D notation	Rel. error (%)
A	$4.2 \times 10^7\,\mathrm{s}^{-1}$	$4.2 \times 10^7\,\mathrm{s}^{-1}$	$4.2 \times 10^7\,\mathrm{s}^{-1}$	10
B	$0.2\,\mathrm{s}^{-1}$	$7 \times 10^{-6}\,\mathrm{cm}^2\mathrm{s}^{-1}$	$3 \times 10^{-12}\,\mathrm{cm}^3 s^{-1}$	30
C	$1.7 \times 10^{-9}\,\mathrm{s}^{-1}$	$3.3 \times 10^{-18}\,\mathrm{cm}^4 s^{-1}$	$4.5 \times 10^{-31}\,\mathrm{cm}^6 s^{-1}$	20
$\mathrm{d}g/\mathrm{d}N$	$4.3 \times 10^{-7}\,\mathrm{cm}^{-1}$	$1.5 \times 10^{-11}\,\mathrm{cm}$	$7.0 \times 10^{-18}\,\mathrm{cm}^2$	10
k_{sat}	4×10^{-8}	$7 \times 10^{-13}\,\mathrm{cm}^2$	$2.2 \times 10^{-17}\,\mathrm{cm}^3$	25
N_{tr}	3.3×10^8	$9.2 \times 10^{12}\,\mathrm{cm}^{-2}$	$2.0 \times 10^{19}\,\mathrm{cm}^{-3}$	10
N_{th}	4.4×10^8	$1.2 \times 10^{13}\,\mathrm{cm}^{-2}$	$2.7 \times 10^{19}\,\mathrm{cm}^{-3}$	10
τ_{th}	2.15 ns	2.15 ns	2.15 ns	10

6.2.4 Recombination Coefficients

Inserting the obtained charge carrier lifetime at threshold of 2.15 ns into Eqs. 6.8 and 6.7 yields a threshold carrier number of $N_{\mathrm{th}} = 4.43 \times 10^8$, which corresponds to a threshold carrier density of $2.73 \times 10^{19}\,\mathrm{cm}^{-3}$ (volume density) or $1.23 \times 10^{13}\,\mathrm{cm}^{-2}$ (area density). The slope of the recombination rate at threshold is $\mathrm{d}r/\mathrm{d}N(N_{\mathrm{th}}) = 1.71\,\mathrm{s}^{-1}$. Solving Eq. 6.2 with these values gives $B = 0.202\,\mathrm{s}^{-1}$ and $C = 1.71 \times 10^{-9}\,\mathrm{s}^{-1}$. In order to transform these coefficients into the material parameters in density notation they are scaled by the total active region volume, which is given by the product of quantum well number and thickness, the ridge width and the cavity length. Finally, an error estimate is obtained by calculating the minimum and maximum values of B and C within the error of the experimentally determined quantities employed in their derivation. Table 6.2 summarizes all device parameters extracted from the analysis of the laser dynamics.

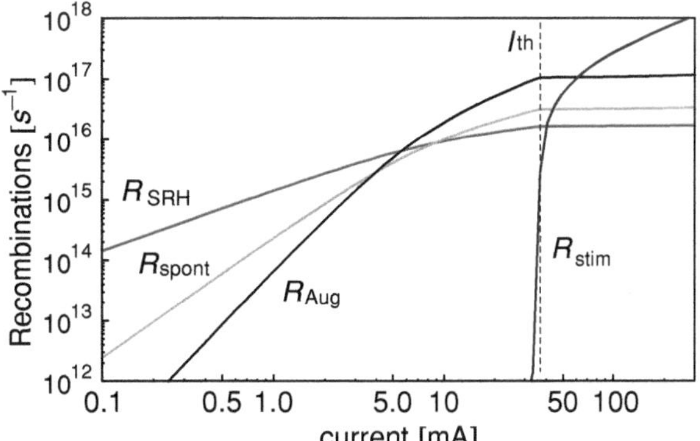

Fig. 6.7 Recombinations due to Shockley-Read-Hall mechanism, spontaneous emission, Auger effect and stimulated emission as functions of pump current. The dashed line marks the threshold current

Apparently, the error of the B-coefficient is rather high for this method. This is due to the fact that the two coefficients B and C are determined from the properties close to and above threshold, where the C-term dominates. An analysis of the internal quantum efficiency could narrow down the range for the spontaneous emission coefficient B, but this is difficult to perform for a laser diode, as stimulated and spontaneous emission cannot easily be separated.

Figure 6.7 shows the individual contributions to the total recombination R from the different mechanisms. As stated above, the third-order recombination term dominates close to threshold. This recombination mechanism is thus highly relevant for laser diodes, because it strongly decreases the charge carrier lifetime, as already pointed out in Refs. [11, 12]. Consequently, the current required to reach a given threshold carrier density increases. For a hypothetical laser diode with identical A and B coefficients and the same N_{th} as the present sample, but $C = 0$, the charge carrier lifetime at threshold would be about 7 ns (compared to 2.15 ns in the actual sample). This means that the presence of the third-order recombination term increases the threshold current by more than a factor of three.

Above threshold the charge carrier density is constant ("clamped"), and so are the non-radiative and spontaneous recombination rates. Any injected current that exceeds threshold is instantly converted into photons via stimulated emission, which becomes the dominant recombination mechanism. For this reason, the injection efficiency can be extracted from the behavior above threshold, as described in Sect. 6.1. The knowledge of the injection efficiency allows a clear differentiation of injection and recombination. It can thus be concluded that the C-coefficient obtained with the present method is not related to charge carrier leakage. The coefficients B and C are determined from the dynamical behavior close to and above threshold, where the charge carrier density in the active region is clamped and a constant injection efficiency can be assumed. This assumption is justified by the linear behavior of both the output power and the square of the oscillation frequency (compare Figs. 6.1 and 6.2). The present findings confirm that there is an intrinsic third-order nonradiative recombination mechanism which is relevant at typical charge carrier densities in

GaN-based light emitters, in accordance with recent excitation-dependent photoluminescence measurements [14].

The numerical value of the C-coefficient obtained from this sample agrees within a factor of 2 with a theoretical study of alloy-disorder- and phonon-assisted Auger scattering by Kioupakis et al., who found a value of about 2×10^{-31} cm^{6}s^{-1} for bulk InGaN with a bandgap of 3 eV [15]. This indicates that the observed mechanism is in fact related to the indirect Auger effect.

6.2.5 Limitations of the ABC-Model

Although the ABC-model is useful to describe the different recombination mechanisms in GaN-based light emitters, it has to be kept in mind that the employed terms are only approximations to the physical recombination rates, the exact form of which are multi-particle integrals in phase space. An exact validity of this model can thus not be assumed. In particular, possible deviations of the spontaneous emission term from quadratic behavior due to a partial screening of internal fields or phase space filling [7] are not considered. Moreover, the C-parameter is not an exact material constant but it should generally depend on the indium content and the width of the quantum wells. Nevertheless, the advantage of the present method is the clear differentiation between injection and recombination, and it is clearly shown that there is a considerable non-radiative recombination term that goes (at least) with N^3 and is not related to charge carrier leakage.

6.3 Gain as Function of Charge Carrier Density

The determination of the functional dependence of the charge carrier lifetime provides a relation between input current and charge carrier density in the active region (see Eq. 6.5). This is very useful for comparing measurements of optical gain in a laser diode to simulations and to verify theoretical models. In the experiment, the measurable quantity related to the pump rate is the current. On the other hand, the input parameter for a gain simulation that relates to the excitation of the system is typically the separation of the quasi-Fermi levels $\mu_e - \mu_h$, which determines the charge carrier density [16]. The charge carrier lifetime thus provides the essential link between experiment and theory.

Figure 6.8a shows the charge carrier density as a function of current, calculated with the recombination coefficients from Table 6.2. As the charge carrier lifetime decreases with increasing occupation of the quantum well, the charge carrier density increases sub-linearly with current. Above threshold, the carrier density is clamped and the excess current is converted directly into photons. By applying the relation between carrier density and current to the optical gain measurements shown in Fig. 6.2, the gain is evaluated as a function of charge carrier density (see Fig. 6.8b).

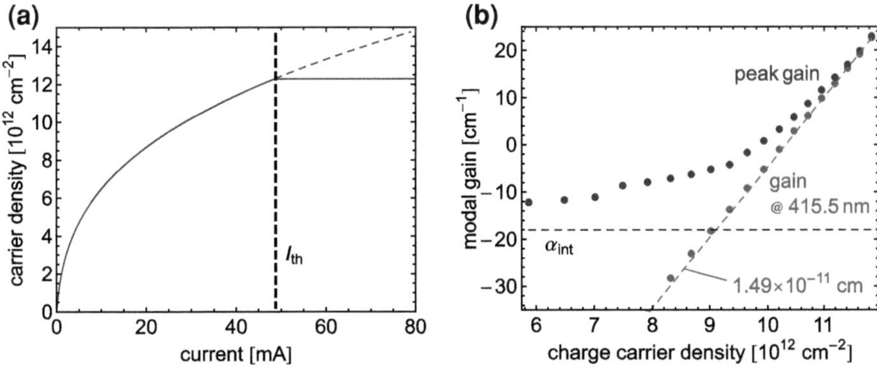

Fig. 6.8 **a** Charge carrier density in each quantum well as a function of pump current. The dashed black line indicates the threshold current, above which the charge carrier density is clamped. **b** Charge carrier density dependence of peak modal gain (*upper points*) and modal gain at the laser wavelength 415.5 nm (*lower points*) with linear fit (*dashed line*). The dashed black line indicates the internal losses α_{int} of the resonator

Fitting the modal gain at the laser wavelength 415.5 nm with a linear function yields a differential gain of 1.49×10^{-11} cm. At a charge carrier density of 9.1×10^{12} cm^{-2}, the linear fit crosses the internal losses, which means that at this carrier density, the quantum wells themselves are transparent for the laser wavelength. This value constitutes the transparency carrier density. Both the differential gain and the transparency carrier density obtained from the gain measurement are in excellent agreement with the parameters determined from the analysis of laser dynamics (see Table 6.2.). This shows that within the single mode rate equation model, the differential gain at the laser wavelength is the relevant parameter, not the differential peak gain, which is generally smaller.

The analysis of the optical gain as a function of charge carrier density also verifies the validity of the linear gain model (see Eq. 2.18). As can be seen from Fig. 6.8b, the gain at the laser wavelength is well fitted by a linear function over a wide range of charge carrier densities. A logarithmic gain model, which is commonly used for laser diodes in other material systems [17] is not appropriate to fit the present data. This indicates that for multi-quantum-well devices with narrow InGaN QWs and at typical threshold carrier densities for GaN-based LDs in the range of 10^{13} cm^{-2}, state filling effects which cause a reduction of the differential gain with increasing carrier density are not pronounced.

References

1. M. Baeumler, M. Kunzer, R. Schmidt, S. Liu, W. Pletschen, P. Schlotter, K. Köhler, U. Kaufmann, J. Wagner, Thermal and non-thermal saturation effects in the output characteristic of UV-to-violet emitting (AlGaIn)N LEDs. Physica Status Solidi A **204**, 1018 (2007)

2. T. Mukai, M. Yamada, S. Nakamura, Characteristics of InGaN-Based UV/Blue/Green/Amber/ Red Light-Emitting Diodes. Jpn. J. Appl. Phys. **38**, 3976 (1999)
3. M. Peter, A. Laubsch, W. Bergbauer, T. Meyer, M. Sabathil, J. Baur, B. Hahn, New developments in green LEDs. Physica Status Solidi A **206**, 1125 (2009)
4. K. Petermann, *Laser Diode Modulation and Noise* (Kluwer Academic Publishers, Dordrecht, 1991)
5. M.-H. Kim, M.F. Schubert, Q. Dai, J.K. Kim, E.F. Schubert, J. Piprek, Y. Park, Origin of efficiency droop in GaN-based light-emitting diodes. Appl. Phys. Lett. **91**(18), 183507 (2007)
6. X. Ni, X. Li, J. Lee, S. Liu, V. Avrutin, U. Ozgür, H. Morkoc, A. Matulionis, T. Paskova, G. Mulholland, K.R. Evans, InGaN staircase electron injector for reduction of electron overflow in InGaN light emitting diodes. Appl. Phys. Lett. **97**(3), 031110 (2010)
7. A. David, M.J. Grundmann, Droop in InGaN light-emitting diodes: a differential carrier lifetime analysis. Appl. Phys. Lett. **96**(10), 103504 (2010)
8. A. Laubsch, M. Sabathil, J. Baur, M. Peter, B. Hahn, High-power and high-efficiency InGaN-based light emitters. IEEE Trans. Electron Devices **57**(1), 79–87 (2010)
9. Y.C. Shen, G.O. Mueller, S. Watanabe, N.F. Gardner, A. Munkholm, M.R. Krames, Auger recombination in InGaN measured by photoluminescence. Appl. Phys. Lett. **91**(14), 141101 (2007)
10. J. Piprek, Efficiency droop in nitride-based light-emitting diodes. Physica Status Solidi A **207**, 2217 (2010)
11. S. Grzanka, P. Perlin, R. Czernecki, L. Marona, M. Bockowski, B. Lucznik, M. Leszczynski, T. Suski, Effect of efficiency "droop" in violet and blue InGaN laser diodes. Appl. Phys. Lett. **95**(7), 071108 (2009)
12. U.T. Schwarz, Emission of biased green quantum wells in time and wavelength domain. Proc. of SPIE 7216, 72161U (2009)
13. S.Y. Karpov, Y.N. Makarov, Dislocation effect on light emission efficiency in gallium nitride. Appl. Phys. Lett. **81**(25), 4721 (2002)
14. A. David, N.F. Gardner, Droop in III-nitrides: comparison of bulk and injection contributions. Appl. Phys. Lett. **97**(19), 193508 (2010)
15. E. Kioupakis, P. Rinke, K.T. Delaney, C.G. Vande Walle, Indirect Auger recombination as a cause of efficiency droop in nitride light-emitting diodes. Appl. Phys. Lett. **98**, 161107 (2011)
16. B. Witzigmann, V. Laino, M. Luisier, U. Schwarz, H. Fischer, G. Feicht, W. Wegscheider, C. Rumbolz, A. Lell, V. Härle, Analysis of temperature-dependent optical gain in GaN-InGaN quantum-well structures. *IEEE Photonics Tech. Letters* **18**(15), 1600–1602 (2006)
17. L.A. Coldren, S.W. Corzine, *Diode Lasers and Photonic Integrated Circuits* (Wiley, New York, 1995)

Chapter 7
Short-Pulse Laser Diodes

Present research on GaN-based laser diodes is focused mainly on the extension of the emission wavelength range, especially towards green emission, and the improvement of maximum output power and efficiency. The principal device concept is the Fabry-Perot-type single-section ridge laser diode. Other device concepts, which have been extensively studied in the group-III-arsenide and -phosphide material systems, are scarcely implemented in the nitrides. However, the advance of GaN-based laser diodes as inexpensive and highly efficient sources of laser light in the visible and near-UV spectral range will enable new applications with special demands, such as single-mode emission, short-pulse generation or fast modulation. Hence, these applications will drive the need to go beyond the basic Fabry-Perot laser concept.

A concept which has been very successfully employed with infrared laser diodes is the multi-section laser diode [1]. Its comprises a ridge laser diode with a cavity that is divided into several electrically insulated sections. Each section can be operated either in forward bias to act as a gain section, or at negative or zero bias to act as a saturable absorber or modulator section. Under appropriate conditions, this system can operate in a self-pulsating regime [2], generating optical pulses on the picosecond scale at a frequency of several GHz. By modulating the absorber section with a frequency that corresponds to the round-trip time in the cavity, active mode-locking can be achieved [3].

This chapter describes the implementation of the multi-section laser diode concept in the group-III-nitrides. The tuneability of the absorber section of GaN-based multi-section LDs is analyzed with respect to the absorption spectrum, the charge carrier lifetime and the impact on the output characteristics. The occurence of self-pulsation is explained by an extended rate equation model that takes into account the charge carrier reservoirs in the gain and absorber sections. Self-pulsating operation of multi-section laser diodes is demonstrated and the dependence of the pulse width and repetition frequency on absorber bias are studied. Additionally, a method to generate single optical pulses is presented.

The samples studied in this chapter are GaN-based multi-section laser diodes grown on c-plane oriented free-standing GaN substrates. Growth and processing were

W. G. Scheibenzuber, *GaN-Based Laser Diodes*, Springer Theses,
DOI: 10.1007/978-3-642-24538-1_7, © Springer-Verlag Berlin Heidelberg 2012

done at the Ecole Polytechnique Federale de Lausanne (EPFL). They have uncoated facets, ridge widths of 2–5 μm and cavity lengths of 600 or 800 μm, with absorber lengths ranging from 25 to 150 μm. The double quantum well active region comprises 3 or 4.5 nm wide InGaN QWs and 12 nm wide unintentionally doped barriers. The waveguide layers on p- and n-side are doped with Mg and Si, respectively, so the applied bias in the absorber drops almost entirely over the ≈40 nm wide active region. The emission wavelengths are around 415 nm.

7.1 GaN-Based Multi-Section Laser Diodes

GaN-based multi-section laser diodes (MS-LDs) are inexpensive and efficient short-pulse lasers which are well-suited for a range of applications. They could be implemented in confocal microscopes for biomedical lifetime imaging, or replace cw laser diodes in optical data storage to reduce feedback noise [4]. At sufficiently high pulse energies in the range of several hundred picojoules, they are suitable for volumetric optical data recording via multiphoton absorption, which would greatly enhance the data capacity of optical discs [5]. For scientific applications that do not require a peak power in the range of kilowatts, GaN-based MS-LDs could also replace frequency-doubled Ti:Sapphire lasers.

Implementing the multi-section laser concept is relatively simple in the group-III-nitrides, owing to the very low lateral conductivity of p-type GaN. To provide adequate electrical insulation between the sections, it is sufficient to define separate contact pads on top of the laser diode and remove a small part of the contact metal stripe, which defines the ridge and also acts as an etch mask during the processing. The separation of the contact metal stripe can be done by electron beam lithography or optical lithography and subsequent dry etching. Thereby, an electrical resistance in the range of 15 kΩ between the sections is achieved [6]. A separation of the underlying semiconductor layers, which would inflict damage to the optical waveguide, is not required. Schematic drawings of multi-section laser diodes in front-absorber and center-absorber configuration are shown in Fig. 7.1. Separate gold pads aside of the ridge are employed to contact the individual sections.

As the absorber section is operated at negative or zero absorber bias, there is no electrical injection of charge carriers into its quantum wells. The properties of the absorber are determined only by the interplay of internal fields and external voltage. In single-section laser diodes the response of the device to a modulation of the pump current is limited by charge carrier dynamics, which occur on a nanosecond time scale. A switching of the absorber of a multi-section laser diode is not limited by charge carrier dynamics. This allows a fast modulation of the output power above threshold [7]. The charge carrier dynamics in single section laser diodes also limit the pulse width that can be achieved by pumping with a short electrical pulse to the width of the relaxation spikes, which is typically around 100 ps (see Sect. 6.1). The implementation of a saturable absorber inside the cavity changes the dynamical behavior of the laser diode and enables the generation of much shorter pulses.

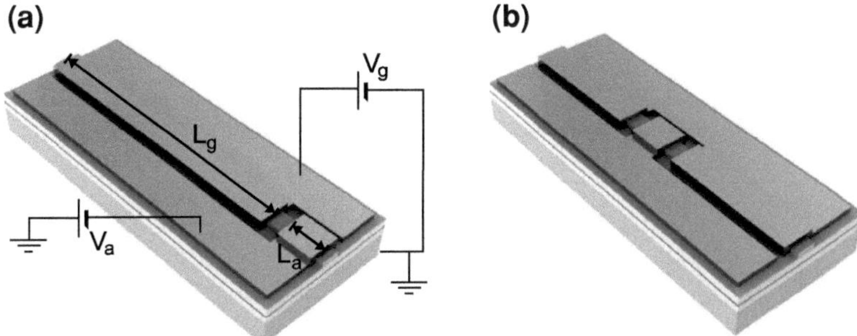

Fig. 7.1 Schematic view (not to scale) of multi-section laser diodes with front absorber (**a**) and center absorber (**b**)

Even though picosecond-pulse generation has been demonstrated also on single-section laser diodes under laboratory conditions [8, 9], the use of multi-section lasers for short-pulse generation appears much more feasible for applications. A pulse width of 15 ps and a peak power of 20 W from a single chip have been demonstrated by Watanabe et al. [10]. By combining their laser diodes with an external cavity setup and a semiconductor optical amplifier, they even achieved 3 ps pulse width and 100 W output power [11]. A physical understanding of the processes that enable the pulse generation will possibly allow an optimization of the heterostructures and a further reduction of the pulse width below one picosecond.

7.2 Tuneability of the Absorber

The occurrence of the self-pulsation regime and the pulse characteristics are determined by the interplay of gain section and absorber section. Absorption spectrum and charge carrier lifetime in the absorber depend on the internal field in the quantum wells of the absorber, which can be tuned via an external bias voltage. In order to optimize the heterostructure for short pulse operation, it is crucial to investigate the properties of the absorber and their tuneability.

7.2.1 Absorption Spectrum

Optical gain spectroscopy is used to determine the absorption spectrum of the InGaN quantum wells in the absorber for different bias voltages. The measurement scheme is shown in Fig. 7.2a, b. In a first step, the multi-section laser diode is treated as a single-section laser by connecting gain and absorber sections and applying a forward voltage. The pump current density j in both sections is set slightly below the laser

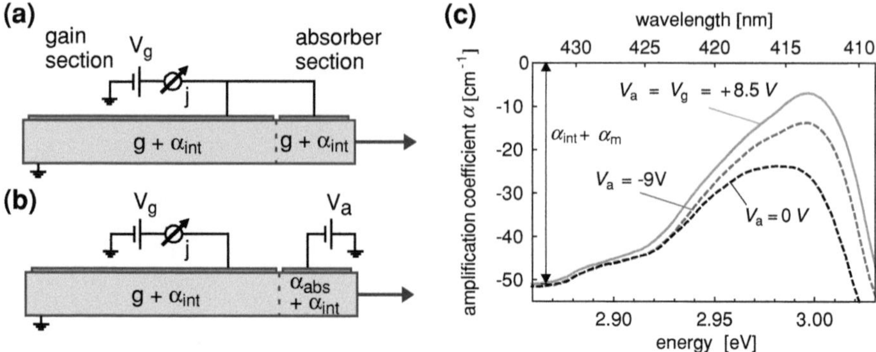

Fig. 7.2 Schematic view of the absorption measurement with gain and absorber section connected (**a**) and different voltages on gain and absorber section (**b**). Measured spectra of the amplification coefficient (**c**) with both sections at equal forward bias (*solid line*) and with 0 V and −9 V (*dashed lines*) absorber bias. The current density in the gain section is the same for all three spectra. The arrow marks the sum of internal losses α_{int} and mirror losses α_m

threshold. As the current density in both sections is identical, they exhibit the same optical gain $g(j, \hbar\omega)$. Using the Hakki-Paoli method, the amplification coefficient α is measured,

$$\alpha(j, \hbar\omega) = -(\alpha_{int} + \alpha_m) + g(j, \hbar\omega), \tag{7.1}$$

from which the optical gain spectrum $g(j, \hbar\omega)$ can be determined. The internal optical losses due to crystal defects and dopant atoms in the waveguide α_{int}, and the mirror losses due to transmission at the laser facets α_m are nearly independent of the photon energy and can be extracted from the long-wavelength tail of the spectrum. Figure 7.2c shows an example measurement on a multi-section laser diode with 3 nm quantum wells, 800 μm cavity and 50 μm front absorber. For the examined sample, the total optical losses $\alpha_{int} + \alpha_m$ are 52 cm^{-1}.

In the second step, the gain section is kept at the same current density while different bias voltages from 7 to −14 V in steps of 1 V are applied to the absorber section. The amplification coefficient is measured for each bias voltage. In a resonator with multiple sections, the Hakki-Paoli method measures the net amplification picked up by the optical mode in one round-trip. This means that the measured amplification coefficient in a multi-section laser diode is an average of the individual sections, weighted with their respective length

$$\alpha(V_a, j, \hbar\omega) = -(\alpha_{int} + \alpha_m) + \frac{L_g}{L_t}g(j, \hbar\omega) - \frac{L_a}{L_t}\alpha_{abs}(V_a, \hbar\omega) \tag{7.2}$$

where L_g, L_a are the lengths of the gain and absorber section, respectively, and $L_t = L_g + L_a$ is the resonator length. V_a is the absorber bias voltage and $\alpha_{abs}(V_a, \hbar\omega)$ is the modal absorption spectrum of the quantum wells in the absorber section.

The influence of the uncontacted region between the two sections is neglected, as it is much smaller than the absorber section. Rearranging Eq. 7.2 gives the absorption spectra of the absorber section

$$\alpha_{abs}(V_a, \hbar\omega) = -\frac{L_t}{L_a}\left(\alpha(V_a, j, \hbar\omega) + (\alpha_{int} + \alpha_m)\right) + \frac{L_g}{L_a}g(j, \hbar\omega) \qquad (7.3)$$

As this measurement technique works below the laser threshold, the number of photo-generated charge carriers in the absorber is negligibly small, which ensures that the absorber is not partially saturated. Also, there is no measurable dependence of the absorption on the gain section current. The systematic error of the Hakki-Paoli measurements is estimated as $0.5\,\mathrm{cm}^{-1}$, which results in an error of $20\,\mathrm{cm}^{-1}$ for the determination of the absorption coefficient due to the factor L_t/L_a in Eq. 7.3.

Resulting absorption spectra for different absorber bias voltages are shown in Fig. 7.3. As expected for the absorption of quantum wells, the measured absorption curves show a pronounced spectral dependence, starting at nearly $0\,\mathrm{cm}^{-1}$ for photon energies way below the effective bandgap of the quantum wells and increasing strongly towards shorter wavelengths. This measurement technique determines the modal absorption coefficient, which is given by the product of the material absorption coefficient of the InGaN quantum wells and the confinement factor Γ of the photon mode in the laser waveguide with the QWs. For the samples with 3 nm quantum wells, $\Gamma = 0.0195$ is calculated using a 2D-waveguide simulation. Dividing the results by the confinement factor gives a maximum value of about $1.3 \times 10^4\,\mathrm{cm}^{-1}$ for the absorption at the laser wavelength 413 nm. A comparison of this value to Ref. [12], which reports a plateau of the quantum well absorption at about $7 \times 10^4\,\mathrm{cm}^{-1}$, reveals that the spectral range examined in the present measurements lies at the low energy tail of the step-like absorption spectrum.

As it can be seen from the inset in Fig. 7.3, the absorption coefficient is also strongly dependent on the applied bias voltage. The modal absorption at the laser wavelength of 413 nm reaches a peak value of $250\,\mathrm{cm}^{-1}$ at a low positive bias voltage of 1 V. A strong kink occurs at 3 V bias, which is due to the onset of a forward current in the absorber. The forward current generates charge carriers in the quantum wells, which decrease the absorption towards transparency. Negative absorber voltages also decrease the absorption towards a minimum of $40\,\mathrm{cm}^{-1}$, which occurs at $-9\,\mathrm{V}$. A further increase of the negative bias voltage then increases the absorption again. A similar bias voltage dependency of the electroabsorption of InGaN quantum well devices has been reported in Refs. [12, 13].

To understand the behavior of the absorption coefficient at negative bias, the measurements are compared to calculations of the absorption coefficient in a laser diode heterostructure. In analogy to the optical gain in a quantum well [14], the absorption is given by

$$\alpha(\hbar\omega) = \sum_{i,f} \Gamma \frac{e^2}{c_0\varepsilon_0\hbar^2\omega n_{gr}d_{QW}} \left(\frac{1}{m_e} + \frac{1}{m_{hh/lh}}\right)^{-1} E_p O_{if} D(\hbar\omega - E_{if}), \quad (7.4)$$

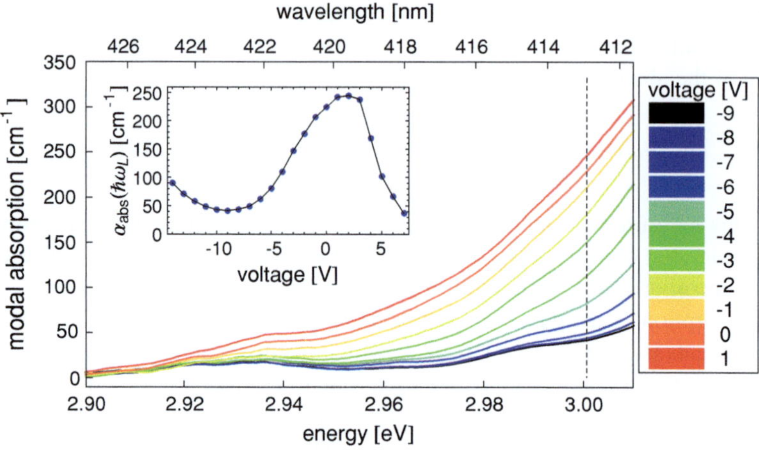

Fig. 7.3 Absorption spectra of the 50 μm absorber at bias voltages from -9 to 1 V. The inset shows the absorption at the laser wavelength of 413 nm as a function of absorber voltage from -14 to 7V

where the sum runs over all bound electron (i) and hole (f) states in the QWs, e is the elementary charge, c_0 is the vacuum speed of light, ε_0 is the vacuum permittivity, n_{gr} is the group refractive index, d_{QW} is the quantum well width, E_{p} is the momentum matrix element energy, m_{e} is the effective electron mass, and $m_{\mathrm{hh/lh}}$ are the effective hole masses for heavy hole (hh) and light hole (lh) bands (split-off holes do not contribute, as their matrix element with TE-polarized light is zero). The parameters required for the simulation, such as effective masses and piezoelectric and spontaneous polarization coefficients, are taken from Ref. [15]. The density of states D is given by a convolution of a step function with a Gaussian, which accounts for homogeneous and inhomogeneous broadening. The Fermi factor ($f_e + f_h - 1$), which appears in the optical gain formula (compare Ref. [14]), is approximated as -1 here, as the charge carrier population of the quantum wells is negligible for negative bias voltages. E_{if} and O_{if} are the transition energy and the squared wave function overlap integral of electron i and hole f, which both depend on the internal electric field in the QW due to the quantum confined Stark effect (QCSE).

The transition energies and wave function overlaps are calculated for a laser diode with two 3 nm wide $\mathrm{In_{0.115}Ga_{0.885}N}$ QWs and 12 nm wide $\mathrm{In_{0.02}Ga_{0.98}N}$ barriers using the SiLENSe package [16]. As shown in Fig. 7.4, the tilt of the quantum wells, and thus the internal electric field, changes with increasing negative bias voltage. The quantum wells become flat at a bias voltage of about -9 V. At this point, the applied voltage compensates the internal field which is caused by the discontinuities of piezoelectric and spontaneous polarization at the material interfaces. The voltage at which this compensation occurs is very sensitive to the design of the active region, namely the width and indium content of the quantum wells and barriers and the doping concentration in the barriers. That is because in strong reverse bias, the voltage drop occurs almost entirely over the undoped regions of the heterostructure. When the

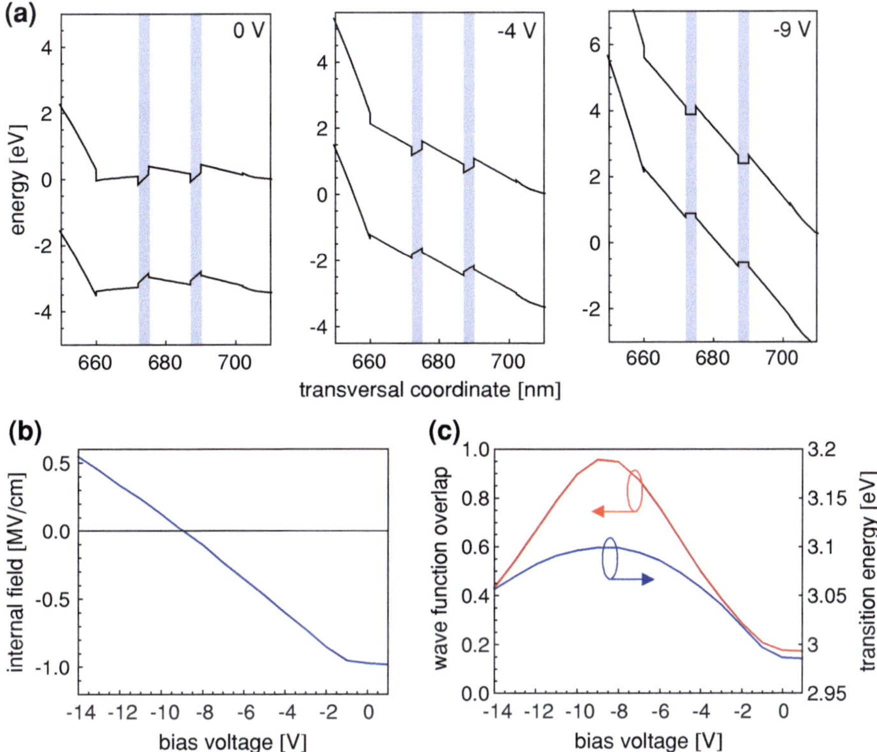

Fig. 7.4 Calculated band edge profiles of the active region (**a**) for 0 V (*left*), −4 V (*center*) and −9V (*right*) bias voltage. The shaded blue regions mark the InGaN quantum wells. Internal field in the quantum wells (**b**), squared wave function overlap (*red*) and transition energy (*blue*) of the electron-heavy hole transition (**c**) as functions of bias voltage

negative bias is decreased, the magnitude of the internal field first increases linearly and then saturates at about −1 V. At low negative bias, the voltage affects mainly the width of the space-charge region on the p-side, so the internal field in the QWs remains nearly constant. The quantum confined Stark effect spatially separates the electrons and holes in the quantum wells and thereby causes a reduction of the wave function overlaps and the transition energies which both scale with the magnitude of the internal field (see Fig. 7.4c).

Absorption spectra are calculated by inserting the obtained overlap integrals O_{if} and transition energies E_{if} into Eq. 7.4. Figure 7.5 shows the results of this calculation for a broadening energy of 50 meV. Even though the wave function overlap, which enters Eq. 7.4 as a factor, increases with increasing negative bias, the shift of the transitions towards higher energies due to the QCSE is the dominating effect, so the absorption is reduced. The simulation reproduces the maximum value of absorption at the laser wavelength (see inset of Fig. 7.5), as well as its spectral dependence.

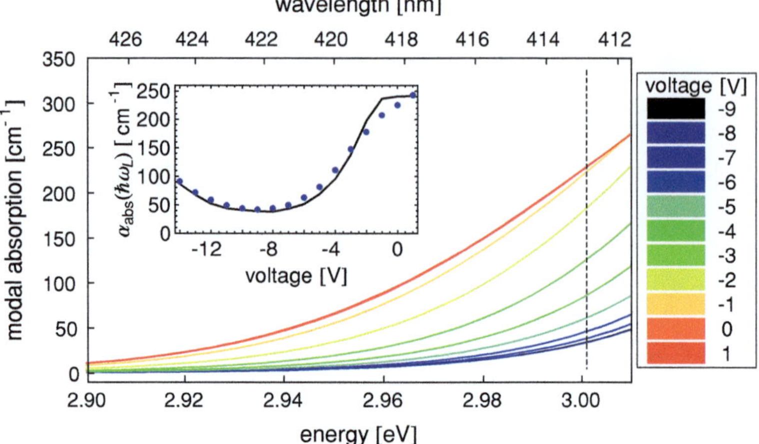

Fig. 7.5 Simulated absorption spectra for bias voltages from -9 to 1 V. The inset shows the calculated absorption at a wavelength of 413 nm as a function of bias voltage from -14 to 1 V (*solid line*) and the experimental data points from Fig. 7.3 (*circles*)

The overall decrease of the absorption with increasing negative bias and the minimum at -9 V, which are observed in the experiment, are also well reproduced.

The observed decrease of the absorption with increasing negative bias is a consequence of the interplay of built-in piezoelectric fields and external bias. Reference [17] reports a different behavior in GaN-based violet multi-section laser diodes, finding an increase of absorption with increasing bias. They study the properties of the absorber using pump probe measurements, where short pulses are injected into the device. However, the behavior of the absorption depends strongly on the details of the epitaxial structure, and the laser diodes studied in Ref. [17] differ significantly from the ones investigated in this work. In particular, they employ Si-doping in the quantum barriers, which strongly influences the internal fields.

7.2.2 Charge Carrier Lifetime

The dynamical behavior of the absorber is determined not only by the absorption coefficient, but also by the charge carrier lifetime in the absorber, which can considerably deviate from the lifetime in the gain section due to the difference in internal field and the escape of charge carriers from the quantum wells in the negatively biased absorber section. The charge carrier lifetime in the absorber is analyzed by applying a variable bias voltage superimposed with a rectangular current pulse to the absorber of a multi-section laser diode with front absorber geometry (see Fig. 7.6a) and measuring the decay of the electroluminescence after a current pulse. The charge carrier lifetime is given by twice the electroluminescence decay time τ_{EL} (compare

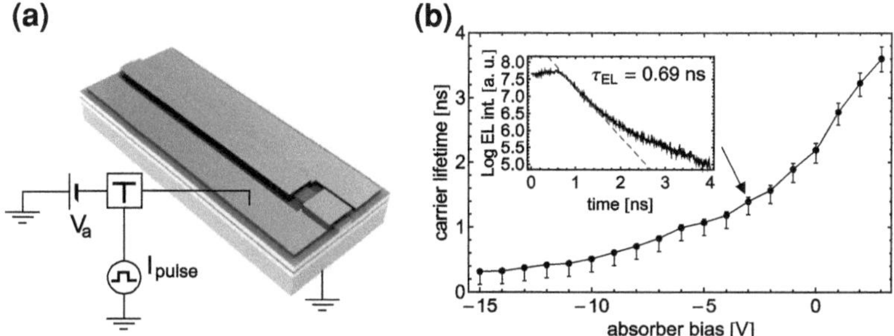

Fig. 7.6 Measurement scheme for the charge carrier lifetime in the absorber (**a**) and charge carrier lifetime τ in the absorber as a function of applied bias voltage (**b**). The asymmetric error bars are due to a systematic overestimation of the decay time because of the finite fall time of the electric pulse. The inset shows an example logarithmic plot of the electroluminescence decay at -3 V bias, where the decay time is obtained from a linear fit at the end of the electric pulse (*dashed line*)

Sect. 6.2.1). For this measurement, a fast electric pulse generator with 100 ps rise/fall time and a streak camera system with a temporal resolution better than 1% of the set time range and a spectral resolution of 0.2 nm are employed. The pulse width is set to 80 ns and the magnitude of the pulse is adjusted to reach a current density of 2 kA/cm^2, which is roughly half the threshold current density of single section lasers from the same epitaxy. The gain section is not contacted during this measurement. Figure 7.6b shows a representative measurement from a multisection laser diode with 4.5 nm quantum wells, 800 μm cavity length and a 100 μm absorber. The charge carrier lifetime decreases drastically with increasing negative bias, from 3.6 ns at $+3$ V to about 400 ps at -15 V.

At a negative bias, the charge carrier lifetime τ_a is composed of the charge carrier escape time τ_{esc}, the radiative lifetime τ_{rad}, and the nonradiative lifetime τ_{nr}, which includes defect-related and Auger recombination:

$$\tau_a = \left(\tau_{esc}^{-1} + \tau_{rad}^{-1} + \tau_{nr}^{-1} \right)^{-1} . \tag{7.5}$$

The reduction of the charge carrier lifetime is due to an increase of radiative recombination and the escape of charge carriers from the quantum wells via thermionic emission and tunneling [18]. The applied bias voltage adds to the field of the p-n junction, which causes a tilt of the band profile and a reduction of the internal fields in the QWs. This leads to an increase of the wave function overlap from about 20% at forward bias to nearly 100% (compare Fig. 7.4) and thereby enhances the radiative recombination. In addition, the tunnel barrier for charge carriers in the QWs is reduced and charge carriers are swept out of the active region, which further decreases their lifetime.

7.3 Absorber and Gain Section Dynamics

Above threshold, the absorber section is pumped optically by photons in the cavity. The increasing occupation of charge carriers in the absorber renders it almost transparent, which causes a rapid change of the overall losses in the cavity and enables the generation of a sharp pulse. This pulse depletes the charge carrier density in the gain section below threshold and subsequently, the photon number in the cavity falls almost to zero (similar to relaxation oscillations). At a low photon number, the optical pumping of the absorber is stopped, rendering it highly absorptive again. If the device properties fulfill certain stability criteria [2], this process repeats itself continuously without decaying, leading to a self-pulsation of the device in a frequency range of several GHz.

The dynamical behavior of a multi-section laser diode above threshold can be simulated using an extension of the rate equation model, which takes into account the charge carrier reservoirs in the gain and absorber section and their coupling to the optical mode [2]. In the area density notation, the rate equations are

$$\frac{dN_g}{dt} = \frac{\eta_{inj}I}{n_{QW}L_g we} - \frac{N_g}{\tau(N_g)} - \frac{L_g}{n_{QW}L_t}g(N_g, S)cS, \tag{7.6}$$

$$\frac{dN_a}{dt} = -\frac{N_a}{\tau_a} + \frac{L_a}{n_{QW}L_t}\alpha(N_a)cS, \tag{7.7}$$

$$\frac{dS}{dt} = \left(\frac{L_g}{L_t}g(N_g, S) - \frac{L_a}{L_t}\alpha(N_a)\right)cS + \beta BN_g^2 - (\alpha_{int} + \alpha_m)cS. \tag{7.8}$$

Here N_g and N_a are the two-dimensional charge carrier densities in each quantum well in the gain and absorber sections, respectively, and S is the area density of photons in the cavity. L_g and L_a are the lengths of the gain and absorber sections, respectively and $L_t = L_g + L_a$. w is the ridge width, n_{QW} is the number of quantum wells, τ_a is the charge carrier lifetime in the absorber, the charge carrier density dependence of which is neglected here, and $\alpha(N_a)$ is the modal absorption of the absorber section. All other quantities appearing here are analog to Eqs. 2.15 and 2.16. As the dependency of α on the charge carrier density is not known, a linear function is assumed,

$$\alpha(N_a) = \alpha_0 (1 - \kappa N_a), \tag{7.9}$$

where α_0 is the unsaturated absorption, and κ is the normalized differential absorption, which is estimated as $1 \times 10^{-12}\,\text{cm}^2$.

This simple simulation model neglects spectral dynamics and thermal effects. It also does not take into account additional saturable absorption effects that arise from charging and uncharging of point defects, which play a role in devices with a rather high defect density. Another source of error is the assumption of a coherent, delocalized photon mode. More sophisticated theoretical models, like the travelling

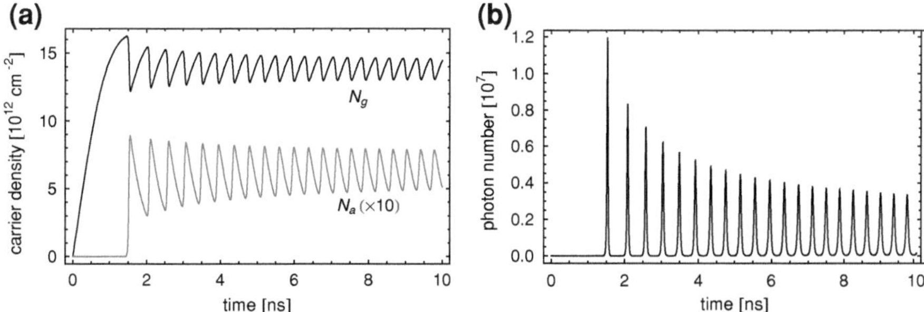

Fig. 7.7 **a** Simulated time evolution of charge carrier density in the gain section N_g and in the absorber section N_a of a multi-section laser diode at the onset of a 180 mA current pulse. **b** Time evolution of photon number in the cavity

wave model [19], treat the propagating pulse in the resonator instead. Therefore, the rate equation model presented here is not suitable for a quantitative comparison to experiments. Nevertheless, it allows to study qualitatively the influence of absorption and charge carrier lifetime in the absorber on the dynamical behavior of the device.

Using the device parameters determined in Chap. 6, he time evolution of photons and charge carriers in a MS-LD with 3 μm ridge, an 800 μm cavity, a 100 μm absorber and the epitaxial design of the samples studied in Chap. 6 is simulated. The results of an example calculation for a pump current of 180 mA, an unsaturated absorption of $300 \, \text{cm}^{-1}$ and a charge carrier lifetime in the absorber of 400 ps are shown in Fig. 7.7. After switching on the current, the gain section fills with charge carriers and a sharp pulse is emitted shortly after the threshold carrier density is reached. This pulse depletes the excess carrier density in the gain section and pumps the absorber optically, so its absorption is partly saturated. After the pulse, the charge carrier density in the absorber drops due to recombination and escape. However, for the given charge carrier lifetime, this drop is not fast enough to entirely deplete the absorber, so there remains an average occupation and the absorption does not return to its initial (unsaturated) value. Therefore, the subsequent pulses experience a smaller change of absorption, which leads to a reduction of their peak power. For a sufficiently high differential absorption and a short charge carrier lifetime in the range of several 100 ps, the output enters a stable self-pulsating operation, whereas a transition to continuous-wave operation occurs if these conditions are not met.

Owing to the fact that the initial spike of the pulse trail in self-pulsating operation is clearly separated from the next pulse, a multi-section laser diode can be used to generate single pulses at arbitrary repetition rates up to 1 GHz. This is particularly useful for applications that rely on a pump-probe technique. Therefore, the gain section is pumped with an electric pulse shorter than 1 ns, so the current is turned off after the first pulse. Mediated by the dynamical behavior of the absorber, the MSLD then provides a considerably shorter optical pulse width than a gain-switched single-section laser diode. For single pulse generation, it is also not relevant whether the self-pulsation of the MSLD would enter a stable regime or not.

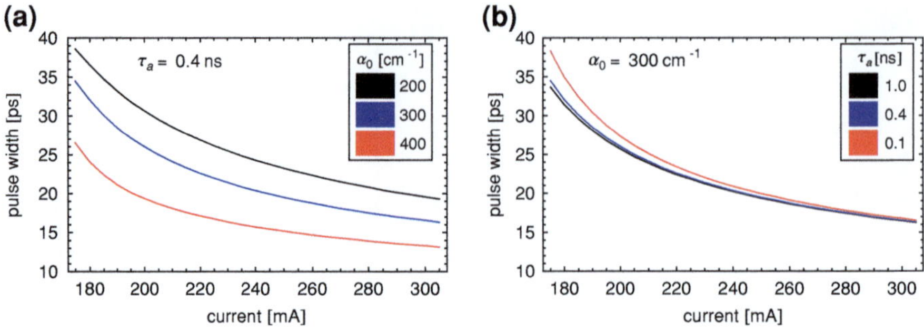

Fig. 7.8 Pulse width as function of pump current for different unsaturated absorption coefficients α_0 at a charge carrier lifetime in the absorber of 0.4 ns (**a**), and for different carrier lifetimes τ_a at $\alpha_0 = 300\,\mathrm{cm}^{-1}$ (**b**)

To determine the influence of absorption and charge carrier lifetime in the absorber, these quantities are varied separately in the simulation. Figure 7.8 shows the calculated pulse width as a function of pump current for different absorption coefficients and lifetimes. The pulse width generally decreases with increasing current, as a high pump rate generates faster dynamics. A plateau in the range of 10–20 ps is reached at high currents. An increase of the absorption coefficient α_0 considerably shifts this plateau to smaller values. On the other hand, the charge carrier lifetime τ_a in the absorber has only a small effect on the width of the optical pulse. It mainly influences the stability of the self-pulsation operation, where stable self-pulsation over a wide range of pump currents is achieved for short values of τ_a. Therefore it is crucial for short pulse generation in multi-section laser diodes to optimize the heterostructure for a high modal absorption in the absorber section.

7.4 Bias-Dependence of the Output Characteristics

Absorption spectrum and charge carrier lifetime in the absorber can be tuned over a wide range via the applied bias. The average output power of the multisection laser diode depends on these quantities, so it also changes with applied bias. A decrease of the absorption coefficient in the absorber reduces the overall losses in the cavity, which have to be compensated by the gain and decreases the threshold. On the other hand, the decrease of the charge carrier lifetime increases the optical pump rate in the absorber which is required to reach partial saturation, leading to an increase of the threshold.

To understand the saturation behavior of the absorber, the averaged output power in constant current operation and the generated photocurrent in the absorber are analyzed as functions of pump current and absorber bias (see Fig. 7.9). At zero bias and low negative bias voltages, a kink in the power–current (P–I) curves is

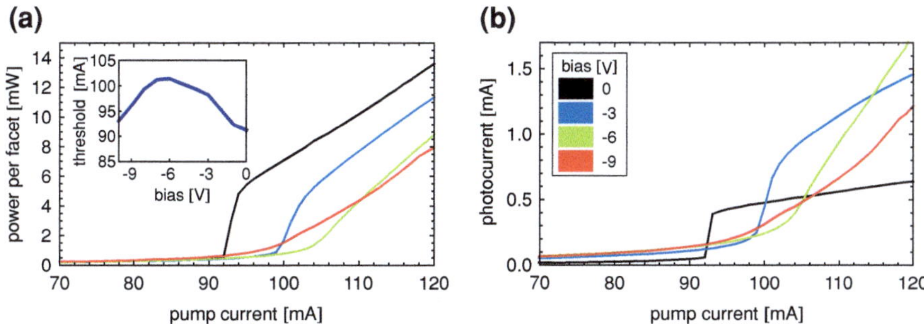

Fig. 7.9 Average output power (**a**) and absorber photocurrent (**b**) as a function of gain section current for different bias voltages in continuous current operation. The inset shows the threshold current as a function of absorber bias voltage

observed, which is characteristic for devices with a saturable absorber [2]. The kink relates to the difference in the absorption coefficient between empty ("off") and partially saturated ("on") absorber state. At low negative bias, the lifetime is too long to allow a depletion of the absorber between two consecutive pulses (compare Sect. 7.3), so there is an average occupation of charge carriers in the absorber above threshold. When the negative bias is increased, the charge carrier lifetime decreases and the average occupation of the absorber with charge carriers is reduced. Therefore, the difference between "on" and "off" state diminishes and the kink disappears. The threshold current increases towards a maximum at -6 V, then decreases again. Apparently, the maximum of the threshold does not coincide with the maximum of the absorption, which occurs around 0 V bias. This is also a consequence of the reduction of the charge carrier lifetime, which reduces the degree of saturation in the absorber and compensates the reduced absorption in empty state. Only when the negative bias is increased beyond -6 V, the reduction of the absorption becomes dominant and the threshold decreases.

The absorber photocurrent I_{ph} relates to the average charge carrier number $\hat{N}_{\mathrm{abs}}$ in the absorber,

$$I_{\mathrm{ph}} = e \frac{\hat{N}_{\mathrm{abs}}}{\tau_{\mathrm{esc}}}, \tag{7.10}$$

where τ_{esc} is the inverse escape rate of charge carriers from the QWs (compare Eq. 7.5). The photocurrent is not just proportional to the output power. It shows a similar kink right at threshold, but it stays almost constant above threshold at zero absorber bias, while at high negative bias it resembles the output power curve. As discussed above, the charge carrier density in the absorber is modulated around an average level, which is determined by the charge carrier lifetime. The behavior of the photocurrent indicates that at low negative bias, the average charge carrier density of the absorber is close to the transparency carrier density, so the absorber is almost completely saturated. Then, a higher photon density in the cavity does not cause a significant increase in photocurrent. At higher negative bias on the other hand, the photocurrent becomes proportional to the output power, which indicates that the average charge carrier density is considerably lower than the transparency carrier density and the absorber is less saturated.

7.5 Self-Pulsation and Single-Pulse Generation

Self-pulsating operation is investigated in GaN-based multi-section laser diodes with different absorber lengths. To characterize the self-pulsation characteristics, they are pumped with 100 ns electric pulses at a repetition frequency of 10 kHz and the output is measured with a streak camera. The pulsation frequencies are determined from 10 ns-long traces each, while the pulse widths are measured at 20 Hz repetition rate in single-shot mode with a time window of 500 ps, which has a temporal resolution of 5 ps. Figure 7.10 shows the time evolution of the output of two multi-section laser diodes with 4.5 nm QWs, 3 μm ridge width and 800 μm cavity length with a 50 μm absorber and a 100 μm absorber at different gain section currents. For the device with the short absorber, stable self-pulsation is observed in a bias range from 0 V to -9 V and up to twice the threshold current, whereas the device with 100 μm absorber exhibits a transition to cw emission at a current of about 190 mA, at 0 V bias. This transition shifts to higher current when the negative bias is increased. The initial pulse of each pulse trail is always considerably higher than the subsequent pulses and slightly red-shifted. This is a consequence of the switching of the absorber from empty state to partly saturated state, as discussed in the previous section (compare Fig. 7.7). The emission spectrum in self-pulsating operation is about 2 nm broad, which is considerably broader than the typical emission of comparable single section devices.

Remarkably, the self-pulsating operation in both devices is stable over a wide range of gain currents and absorber bias voltages, even though Eqs. 7.6–7.8 predict continuous-wave operation for the measured values of modal absorption and charge carrier lifetime. However, this simple model does not include the charging and uncharging of point defects, which is known to cause self-oscillations in single section laser diodes [1, 20]. It is likely that this effect stabilizes the self-pulsation in the present samples.

As Fig. 7.11 shows, the oscillation frequency of both devices increases monotonically within a range of 1.5 to 5 GHz with increasing current. For the 50 μm absorber, a comparison of the dependency of the oscillation frequencies on pump current and absorber bias with the average output power (see Fig. 7.9) reveals that the self-pulsation frequency goes as the squareroot of the output power, just like the frequency of relaxation oscillations (compare Sect. 6.1). Therefore, the occurrence of self-pulsation can be attributed to a stabilization of relaxation oscillations by the saturable absorber. At low currents, the device with the 100 μm absorber behaves similarly, but a transition to cw operation with initial relaxation oscillations occurs at higher current, and the frequency of these relaxation oscillations depends rather

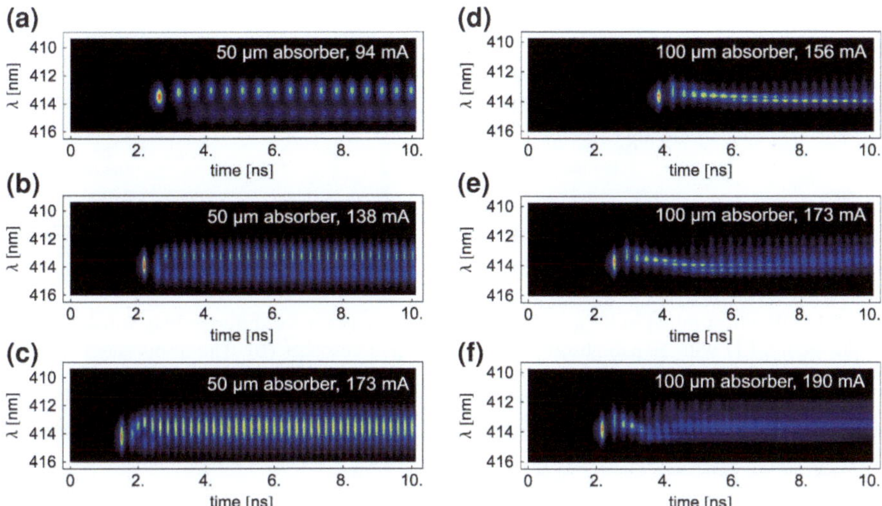

Fig. 7.10 Example streak camera traces of self-pulsation at 0 V absorber bias from a multi-section laser diode with 50 μm absorber at 94 mA (**a**), 138 mA (**b**), 173 mA (**c**) gain section current and from a sample with 100 μm absorber at 156 mA (**d**), 173 mA (**e**), 190 mA (**f**)

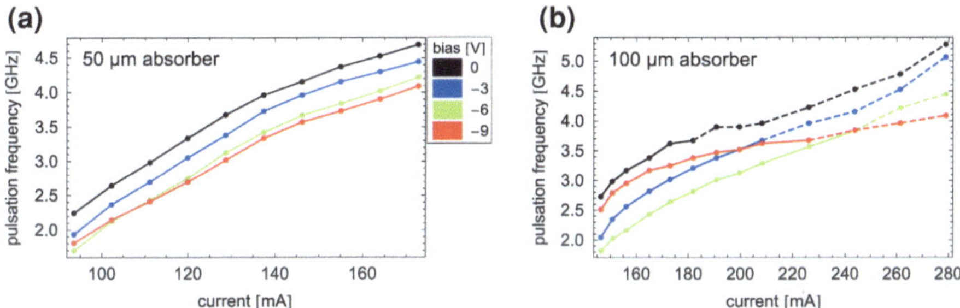

Fig. 7.11 Oscillation frequency as a function of pump current for different bias voltages for the multi-section LD with 50 μm absorber (**a**) and 100 μm absorber (**b**). The *dashed lines* in **b** indicate the current range where the oscillations become unstable

linearly on the pump current. The dependency of the self-pulsation frequency on pump current observed in this work differs from the behavior reported in Refs. [6, 21], where an almost constant pulsation frequency around 1 GHz was obtained. However, it is known from other material systems that the pulsation frequency depends on the active region design, in particular on the magnitude of doping in the quantum wells and barriers [22].

The width of the initial pulse decreases as a function of current, reaching a plateau of 18 ps at 173 mA and zero absorber bias for the 50 μm absorber and 21 ps at 283 mA for the 100 μm absorber (see Fig. 7.12). With increasing negative bias the pulses become longer, indicating that the pulse width depends in fact mainly on

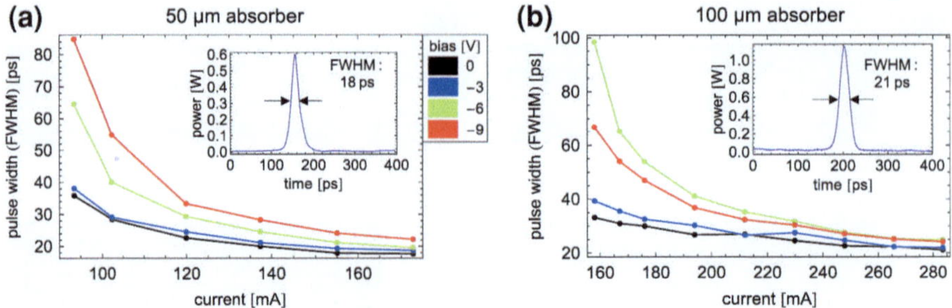

Fig. 7.12 Width of initial pulse as a function of pump current for different bias voltages for the multi-section LD with 50 μm absorber (**a**) and 100 μm absorber (**b**). The insets show the pulse shape at 173 mA (**a**) and 283 mA (**b**), respectively, and 0 V bias voltage

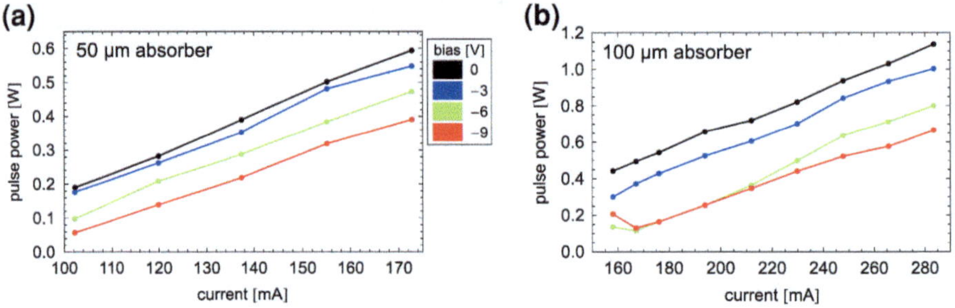

Fig. 7.13 Peak power of single pulse as a function of pump current for different bias voltages for the multi-section LD with 50 μm absorber (**a**) and 100 μm absorber (**b**)

the value of the absorption coefficient in the absorber section. This finding is in qualitative agreement with the simulation results presented in Sect. 7.3. Comparing the absorber lengths, one finds that the device with the longer absorber exhibits a longer pulse width even at much higher current. Therefore, one can conclude that a short absorber with high modal absorption is desirable for generating short pulses.

Single pulses are generated without a loss of peak intensity simply by reducing the duration of the electric excitation pulse to a value between 0.5 and 2 ns, depending on the pump current. The peak power of the pulses is then determined by measuring the average output power and dividing by the product of the repetition frequency (10 kHz) and the pulse width (see Fig. 7.13). The maximum peak power is 0.60 ± 0.05 W at 173 mA pump current for the 50 μm absorber and 1.1 ± 0.1 W at 283 mA pump current for the 50 μm absorber, both at 0 V absorber bias voltage. The corresponding maximum pulse energies for both devices are 11 ± 1 pJ and 23 ± 2 pJ, for the 50 and 100 μm absorber, respectively. Even though the device with the longer absorber does not exhibit stable self-pulsation at high currents, it is suitable for single pulse generation, as the first spike has a proper shape (see inset of Fig. 7.12b). The maximum achieved peak power is almost twice the value measured for the device with short absorber, and the pulse width is only 3 ps longer. A longer absorber thus provides a higher peak power, at the cost of a slightly increased pulse width.

References

1. C. Mirasso, G. Van Tartwijk, E. Hernandez-Garcia, D. Lenstra, S. Lynch, P. Landais, P. Phelan, J. O'Gorman, M. San Miguel, W. Elsasser, Self-pulsating semiconductor lasers: theory and experiment. IEEE J. Quantum Electron. **35**(5), 764–770 (1999)
2. M. Ueno, R. Lang, Conditions for self-sustained pulsation and bistability in semiconductor lasers. J. Appl. Phys. **58**(4), 1689–1692 (1985)
3. P. Acedo, H. Lamela, S. Garidel, C. Roda, J. Vilcot, G. Carpintero, I. White, K. Williams, M. Thompson, W. Li et al., Spectral characterisation of monolithic modelocked lasers for mm-wave generation and signal processing. Electron. Lett. **42**(16), 928–929 (2006)
4. Y. Kawaguchi, Y. Tani, P.O. Vaccaro, S. Ito, H. Kawanishi, Electric field induced carrier sweep-out in tandem InGaN multi-quantum-well self-pulsating laser diodes. Jpn. J. Appl. Phys. **50**(2), 020209 (2011)
5. S. Tashiro, Y. Takemoto, H. Yamatsu, T. Miura, G. Fujita, T. Iwamura, D. Ueda, H. Uchiyama, K. Yun, M. Kuramoto, T. Miyajima, M. Ikeda, H. Yokoyama, Volumetric optical recording using a 400 nm all-semiconductor picosecond laser. Appl. Phys. Express **3**(10), 102501 (2010)
6. T. Miyajima, H. Watanabe, M. Ikeda, H. Yokoyama, Picosecond optical pulse generation from self-pulsating bisectional GaN-based blue-violet laser diodes. Appl. Phys. Lett. **94**, 161103 (2009)
7. M. Kneissl, T.L. Paoli, P. Kiesel, D.W. Treat, M. Teepe, N. Miyashita, N.M. Johnson, Two-section InGaN multiple-quantum-well laser diode with integrated electroabsorption modulator. Appl. Phys. Lett. **80**(18), 3283 (2002)
8. S. Kono, T. Oki, T. Miyajima, M. Ikeda, 12 W peak-power 10 ps duration optical pulse generation by gain switching of a single-transverse-mode GaInN blue laser diode. Appl. Phys. Lett. **93**, 131113 (2008)
9. M. Kuramoto, T. Oki, T. Sugahara, S. Kono, M. Ikeda, H. Yokoyama, Enormously high-peak-power optical pulse generation from a single-transverse-mode GaInN blue-violet laser diode. Appl. Phys. Lett. **96**, 051102 (2010)
10. H. Watanabe, M. Kuramoto, S. Kono, M. Ikeda, H. Yokoyama, Blue-violet bow-tie self-pulsating laser diode with a peak power of 20W and a pulse energy of 310pJ. Appl. Phys. Express **3**, 3–5 (2010)
11. R. Koda, T. Oki, T. Miyajima, H. Watanabe, M. Kuramoto, M. Ikeda, H. Yokoyama, 100 W peak-power 1 GHz repetition picoseconds optical pulse generation using blue-violet GaInN diode laser mode-locked oscillator and optical amplifier. Appl. Phys. Lett. **97**, 021101 (2010)
12. F. Renner, P. Kiesel, G.H. Döhler, M. Kneissl, C.G. Van de Walle, N.M. Johnson, Quantitative analysis of the polarization fields and absorption changes in InGaN/GaN quantum wells with electroabsorption spectroscopy. Appl. Phys. Lett. **81**(3), 490 (2002)
13. P. Kiesel, F. Renner, M. Kneissl, N. Johnson, G. Döhler, Electroabsorption spectroscopy— direct determination of the strong piezoelectric field in InGaN/GaN heterostructure diodes. Physica Status Solidi A **188**(1), 131–134 (2001)
14. W.W. Chow, S.W. Koch, *Semiconductor-Laser Fundamentals* (Springer, Berlin, 1998)
15. I. Vurgaftman, J. Meyer, in *Electron Bandstructure Parameters* ed. by J. Piprek. Nitride Semiconductor Devices: Principles and Simulations, chap. 2 (Wiley VCH, Weinheim, 2007), pp. 13–18
16. STR Group Ltd., Simulator of Light Emitters based on Nitride Semiconductors (SiLENSe). http://www.semitech.us/products/SiLENS
17. T. Miyajima, S. Kono, H. Watanabe, T. Oki, R. Koda, M. Kuramoto, M. Ikeda, H. Yokoyama, Saturable absorbing dynamics of GaInN multiquantum well structures. Appl. Phys. Lett. **98**(17), 171904 (2011)
18. U.T. Schwarz, H. Braun, K. Kojima, Y. Kawakami, S. Nagahama, T. Mukai, Interplay of built-in potential and piezoelectric field on carrier recombination in green light emitting InGaN quantum wells. Appl. Phys. Lett. **91**(12), 123503 (2007)

19. Y.L. Wong, J.E. Carroll, A travelling-wave rate equation analysis for semiconductor lasers. Solid-State Electron. **30**(1), 13–19 (1987)
20. S. Nakamura, M. Senoh, S.-I. Nagahama, N. Iwasa, T. Yamada, T. Matsushita, H. Kiyoku, Y. Sugimoto, T. Kozaki, H. Umemoto, M. Sano, K. Chocho, InGaN/GaN/AlGaN-based laser diodes with modulation-doped strained-layer superlattices. Jpn. J. Appl. Phys. **36**(12A), 1568–1571 (1997)
21. H. Watanabe, T. Miyajima, M. Kuramoto, M. Ikeda, H. Yokoyama, 10-W peak-power picosecond optical pulse generation from a triple section blue-violet self-pulsating laser diode. Appl. Phys. Express **3**(5), 052701 (2010)
22. T. Tanaka, T. Kajimura, Frequency control of self-sustained pulsating laser diodes by uniform impurity doping into multiple-quantum-well structures. IEEE Photonics Tech. Lett. **10**(1), 48–50 (1998)

Chapter 8
Summary and Conclusions

The present work treats various aspects of current research on GaN-based laser diodes. It focuses on efforts and possible ways to increase the emission wavelength and realize true-green laser diodes for projection and display applications. Furthermore, it covers the realization of multi-section laser diodes on GaN, a device concept for the generation of short pulses, which is successfully adapted from infrared laser diodes. Spectral and temporal characterization methods are applied to various laser diode samples to investigate the influence of the heterostructure design on the device parameters. Where appropriate, the experimental results are compared to theoretical models to gain insight on the physical processes in the device and to separate individual contributions to certain phenomena. Although the models presented here are simplified and in most cases not suitable for quantitative predictions, their strength lies in their versatility and compactness, providing a qualitative understanding of trends at a comparably low numerical effort.

By analyzing the influence of temperature on the emission spectra below and above threshold, a method is developed to precisely determine the thermal resistance of GaN-based laser diodes, which typically lies in the range of 20 to 40 K/W for LDs on GaN substrate. The knowledge of this quantity allows to suppress thermal effects in other experiments by appropriate cooling of the heat sink, depending on the dissipated power. This method is also used to monitor the time evolution of the internal temperature on a nanosecond time scale using time-resolved spectroscopy. Thereby, two coupled subsystems of the laser diode are identified, which heat up on different time scales. While the charge carrier plasma of the studied sample heats up with a time constant of 6 ns, the crystal lattice thermalizes in 0.4 µs. This means that even in short-pulsed operation and at low duty cycle, heating effects cannot be completely avoided in GaN-based laser diodes.

Optical gain spectra of highly optimized multi-quantum-well (Al,In)GaN laser diodes with different indium contents are compared, and clear trends are observed regarding differential gain and spectral width. This work presents the first electrically pumped optical gain measurement on a GaN-based true green laser diode. The observed reduction of the optical gain for increasing emission wavelength is

W. G. Scheibenzuber, *GaN-Based Laser Diodes*, Springer Theses,
DOI: 10.1007/978-3-642-24538-1_8, © Springer-Verlag Berlin Heidelberg 2012

attributed to the increase in indium content, which causes on the one hand an inferior material quality and on the other hand a reduction of the electron-hole wave function overlap due to the strong piezoelectric fields. From the separation of the longitudinal modes and the current dependent shift of the spectra, the group refractive index and the charge carrier induced refractive index change are calculated. The observed reduction of the group refractive index from 3.27 for violet to 2.90 for blue and 2.75 for green reflects the dependence of the refractive indices of the waveguide materials on the wavelength. The antiguiding factor is calculated from the quotient of refractive index change and differential gain. It has values of 3.4, 4.1 and 4.3 for the violet, blue and green laser diode, respectively. Interestingly, the antiguiding factor is very similar for all three emission wavelengths even though optical gain and refractive index change are strongly affected by the different magnitudes of the quantum confined Stark effect.

Based on $k \cdot p$-simulations of band structure and optical gain, possible improvements for green laser diodes are investigated that arise from the growth on crystal planes which are inclined to the c-plane. This work demonstrates how to implement the influence of birefringence, which occurs in wurzite group-III-nitrides, in gain calculations for quantum wells on such semipolar crystal planes. Depending on the orientation of the waveguide relative to the crystal, the optical eigenmodes are polarized along the TE/TM directions or along the extraordinary and ordinary directions of the birefringent crystal. Calculated optical gain spectra for the different eigenmodes are compared for the $(11\bar{2}2)$- and the $(20\bar{2}1)$-plane, two planes on which laser operation has already been demonstrated in various publications. For the $(20\bar{2}1)$-plane, the optical gain is by a factor of six higher than for a c-plane laser diode with similar emission wavelength, but only for the waveguide orientation which has a TE-polarized optical eigenmode. The optical gain in green emitting quantum wells on the $(11\bar{2}2)$-plane is three times higher than on the c-plane. On the $(11\bar{2}2)$-plane, a switching of the dominant optical polarization occurs at an indium content of about 28%, which results in similar optical gain for both waveguide orientations. Owing to this effect, a waveguide orientation with a low index cleavage plane can be chosen, which facilitates the fabrication of smooth mirror facets.

Using time-resolved spectroscopy, the dynamics of charge carriers and photons in GaN-based laser diodes are analyzed. From a comparison of the experimentally observed turn-on delay and the relaxation oscillations of the laser diode to a rate equation model, dynamical parameters of the device are extracted such as the differential gain per charge carrier density, the charge carrier lifetime at threshold and the gain saturation parameter. Furthermore, this work demonstrates a method to determine the individual charge carrier recombination coefficients by combining measurements of optical gain and laser dynamics. The main advantage of this method is that the effects of charge carrier leakage and higher order recombination can be clearly differentiated. A third order recombination term of $C = 4.5 \pm 0.9\,\mathrm{cm^6 s^{-1}}$ is obtained, which is not related to charge carrier leakage. This value is within a factor of 2 in agreement with a theoretical study of indirect Auger recombination for bulk InGaN, which is a strong indication that the indirect Auger effect plays a major role in the drop of the efficiency with increasing current density experienced in group-III-nitride light emitters.

The concepts used to investigate the properties of continuous-wave laser diodes are extended to analyze multi-section laser diodes for short pulse generation. An experimental method is presented which allows the measurement of the absorption spectrum of the quantum wells in the absorber section depending on the applied bias voltage. Here, a special feature of the group-III-nitride material system is revealed, as the absorption first decreases with increasing negative bias, then reaches a minimum and increases again. Using a band-profile simulation, this behavior is attributed to the piezoelectric field, which causes a redshift of the absorption edge relative to the laser emission wavelength via the quantum confined Stark effect. This internal field is increasingly compensated by the external bias, so the absorption edge is blue-shifted. It is also shown that the charge carrier lifetime in the absorber decreases to few hundred picoseconds at high negative bias due to the escape of charge carriers from the quantum well and an increase in radiative recombination due to the untilting of the quantum wells. An extended rate equation model reveals that the major influence on the optical pulse width is the absorption coefficient of the absorber. Therefore, the laser heterostructure has to be optimized for a high absorption in the absorber section in order to achieve pulse widths below 10 ps. The self-pulsating operation of the devices is characterized regarding pulse width, repetition frequency and peak power as functions of pump current and absorber bias. Here, a square-root like increase of the frequency from 1.5 to 5 GHz with increasing current is found, which is similar to the current dependence of the relaxation frequency in single-section laser diodes. Therefore, the self-pulsation is attributed to a stabilization of relaxation oscillations by the dynamical behavior of the saturable absorber. A minimum pulse width of 21 ps and a peak power of 1.1 W are demonstrated on a device with 100 μm long absorber. This corresponds to a pulse energy of 23 pJ. An increasing absorber bias voltage is found to decrease the peak power and increase the pulse width, owing to the reduction of the absorption coefficient of the absorber section.

In summary, this work provides insight in various aspects of the device physics in GaN-based laser diodes. Sophisticated experimental and theoretical methods are employed to determine a wide range of device parameters. The knowledge of these parameters allows a specific optimization of heterostructures towards the requirements of the targeted applications. A further understanding of the physics of GaN-based laser diodes will pave the way to a wider range of emission wavelengths, higher output powers and efficiencies and the realization of special device concepts, such as multi-section laser diodes, distributed Bragg reflector lasers and tapered amplifiers.

Appendix
Numerical Methods

A.1 Numerical Solution of the Quantum Well Schrödinger Equation

Wave equations like Eq. (4.13) can be solved numerically by expanding the eigenfunctions as finite Fourier series and thereby transforming the problem to a matrix eigenvalue equation. To do so, one has to impose periodic boundary conditions on a reference length L along the z-direction. The choice of this length affects the basis functions of the expansion, and it is useful to choose a multiple of the quantum well thickness, for example $L = 8d$. The following ansatz is used for the eigenfunctions:

$$\vec{\psi}(z) = \sum_{n=0}^{N} \frac{1}{\sqrt{L}} \vec{c}_n \exp(iq_n z), \tag{A.1}$$

with $q_n = \frac{\pi}{L}(N - 2n)$. N is the number of basis functions in which to expand and it has to be even to include $q = 0$. These eigenfunctions are exact only in the limit $N \to \infty$, but it is found that for the ground state and the lower excited states the solutions converge rapidly and acceptable accuracy can already be achieved with $N = 100$. Plane waves are chosen as basis functions because they reduce the matrix operator $\mathbf{H_v}(k_x, k_y, -i\frac{d}{dz})$ to the bulk Hamiltonian $\mathbf{H_v}(k_x, k_y, q_n)$. Inserting this ansatz into Eq. (4.13), multiplying by $\frac{1}{\sqrt{L}}\exp(-iq_m z)$ and integrating over L yields

$$((\mathbf{H_h}(k_x, k_y, q_n))_{ij}\delta_{mn} + V_{mn}\delta_{ij})(c_n)_j = E(c_m)_i, \tag{A.2}$$

where δ_{mn}, δ_{ij} are Kronecker deltas and V_{mn} is the Fourier transformation of the potential,

$$V_{mn} = \frac{1}{L} \int_{-L/2}^{L/2} dz \exp(-i(q_m - q_n)z) V_{VB}(z). \tag{A.3}$$

W. G. Scheibenzuber, *GaN-Based Laser Diodes*, Springer Theses,
DOI: 10.1007/978-3-642-24538-1, © Springer-Verlag Berlin Heidelberg 2012

Sum convention is used here and $i, j = 1$ to 6 number the components of the vectors $\vec{c}_n$ and $\vec{c}_m$, whereas the indices $n, m = 1$ to N number the basis functions of the Fourier expansion (A.1). To treat this problem numerically, the indices j,n and i,m are transformed to one index,

$$\alpha = 6n + j, \alpha' = 6m + i, \tag{A.4}$$

where α, α' now go from 1 to $6(N + 1)$. With this transformation, Eq. (A.2) becomes a simple matrix eigenvalue equation:

$$\mathcal{H}_{\alpha\alpha'} c_{\alpha'} = E c_\alpha, \tag{A.5}$$

with $\mathcal{H}_{(6n+j),(6m+i)} = (\mathbf{H_v}(k_x, k_y, q_n))_{ij}\delta_{mn} + V_{mn}\delta_{ij}$. This equation can be solved numerically with standard diagonalization algorithms, yielding both the energy eigenvalues and the corresponding eigenfunctions for k_x, k_y as vectors with $6(N + 1)$ components, which can be decomposed into the wave function's Fourier coeffecient vectors $\vec{c}_n$ by inverting the index transformation (A.4). This method is also applicable for the conduction band Schrödinger equation, simplified by the fact that its eigenfunctions are scalar, so an index transformation like (A.4) is not necessary.

A.2 4×4 Transfer Matrix Waveguide Calculation

The 4×4 transfer matrix method calculates the wavefront in an anisotropic planar layer waveguide for a wave that propagates in y-direction with layers perpendicular to the z-direction. The Maxwell equations for such a structure are:

$$\nabla \cdot (\varepsilon\vec{E}) = 0 \tag{A.6}$$

$$\nabla \cdot \vec{B} = 0 \tag{A.7}$$

$$\nabla \times \vec{E} = i\omega\vec{B} \tag{A.8}$$

$$\nabla \times \vec{B} = -i\frac{\omega}{c_0^2}\varepsilon\vec{E} \tag{A.9}$$

Here, time derivatives are replaced by $-i\omega$, ε is the nondiagonal dielectric tensor (with components ε_{ij}) that depends on z and c_0 is the speed of light in vacuum. For this method, it is assumed that the electric and magnetic fields do not depend on the x-coordinate, which is true for a wave propagating in the y-direction in an infinite film. This simplification is justified if the lateral extensions of the planar waveguide are much bigger than the transversal ones. The wave propagates in y-direction, so the y-dependency is simply $\exp(iyn_{\mathrm{eff}}\omega/c_0)$ and all y-derivatives are replaced by $in_{\mathrm{eff}}\omega/c_0$, where n_{eff} is the effective index of refraction. This effective index is to be determined by the calculation. Rearranging Eqs. (A.8) and (A.9) gives for the components of the fields:

$$\frac{d}{dz}\begin{pmatrix} E_y \\ c_0 B_x \\ E_x \\ c_0 B_y \end{pmatrix} = i\frac{\omega}{c_0}\Delta\begin{pmatrix} E_y \\ c_0 B_x \\ E_x \\ c_0 B_y \end{pmatrix} \tag{A.10}$$

$$c_0 B_z = n_{\text{eff}} E_x \tag{A.11}$$

$$E_z = -\frac{\varepsilon_{zy}}{\varepsilon_{zz}} E_y - \frac{\varepsilon_{zx}}{\varepsilon_{zz}} E_x - \frac{n_{\text{eff}}}{c_0} \varepsilon_{zz} B_x \tag{A.12}$$

$$\Delta = \begin{pmatrix} -n_{\text{eff}}\frac{\varepsilon_{zy}}{\varepsilon_{zz}} & 1 - \frac{n_{\text{eff}}^2}{\varepsilon_{zz}} & -n_{\text{eff}}\frac{\varepsilon_{zx}}{\varepsilon_{zz}} & 0 \\ \varepsilon_{yy} - \frac{\varepsilon_{yz}\varepsilon_{zy}}{\varepsilon_{zz}} & -n_{\text{eff}}\frac{\varepsilon_{yz}}{\varepsilon_{zz}} & \varepsilon_{yx} - \frac{\varepsilon_{yz}\varepsilon_{zx}}{\varepsilon_{zz}} & 0 \\ 0 & 0 & 0 & -1 \\ -\varepsilon_{xy} + \frac{\varepsilon_{xz}\varepsilon_{zy}}{\varepsilon_{zz}} & n_{\text{eff}}\frac{\varepsilon_{xz}}{\varepsilon_{zz}} & n_{\text{eff}}^2 - \varepsilon_{xx} + \frac{\varepsilon_{xz}\varepsilon_{zx}}{\varepsilon_{zz}} & 0 \end{pmatrix} \tag{A.13}$$

Formal integration of Eq. (A.10) leads to:

$$\vec{\phi}(z_{j+1}) = \exp(i\omega(z_{j+1} - z_j)\Delta)\vec{\phi}(z_j)$$
$$=: T_L(z_{j+1} - z_j)\vec{\phi}(z_j) \tag{A.14}$$

$$\vec{\phi} = \begin{pmatrix} E_y \\ c_0 B_x \\ E_x \\ c_0 B_y \end{pmatrix} \tag{A.15}$$

T_L is the transfer matrix for one layer. Multiplying the transfer matrices for all layers gives the T matrix, which connects the bottom and the top of the structure. For a guided mode, all fields must vanish for $z \to \pm\infty$. Assuming that the outside of the structure is vacuum, the matrix Δ for the outside takes the simple form

$$\Delta = \begin{pmatrix} 0 & 1 - n_{\text{eff}}^2 & 0 & 0 \\ 1 & 0 & 0 & 0 \\ 0 & 0 & 0 & -1 \\ 0 & 0 & n_{\text{eff}}^2 - 1 & 0 \end{pmatrix} \tag{A.16}$$

This matrix has two twice degenerate eigenvalues $\pm i\beta$ with $\beta = \sqrt{1 - n_{\text{eff}}^2}$. There are thus two $\vec{\phi}$ vectors that rise exponentially for $z \to \infty$ and decay for $z \to -\infty$, and two others that behave vice versa. The waveguiding condition now states that a guided mode must consist only of vectors that decay exponentially on the outside of the structure. An eigenvector to $-i\beta$ on the bottom should thus become an eigenvector to $+i\beta$ when transformed with the T-matrix, which gives the equation:

$$T \left(a \begin{pmatrix} 0 \\ 0 \\ i \\ -\beta \end{pmatrix} + b \begin{pmatrix} \beta \\ i \\ 0 \\ 0 \end{pmatrix} \right) = c \begin{pmatrix} 0 \\ 0 \\ i \\ \beta \end{pmatrix} + d \begin{pmatrix} -\beta \\ i \\ 0 \\ 0 \end{pmatrix} \tag{A.17}$$

$$\underbrace{}_{\text{eigenvectors to } -i\beta \text{ in vacuum}} \quad \underbrace{}_{\text{eigenvectors to } i\beta \text{ in vacuum}}$$

This is a homogeneous linear equation system for the coefficients a, b, c, d. A non-zero solution exists if the determinant vanishes. The waveguiding condition in its explicit form is then:

$$(T_{43} + i\beta(T_{44} + T_{33}) - \beta^2 T_{34})(-T_{12} + i\beta(T_{11} + T_{22}) + \beta^2 T_{21})$$
$$+ (T_{42} + i\beta(T_{32} - T_{41}) + \beta^2 T_{31})(T_{13} + i\beta(T_{14} - T_{23}) + \beta^2 T_{24}) = 0 \tag{A.18}$$

The T_{ij} are functions of n_{eff}, so Eq. (A.18) allows the calculation of n_{eff} such that a guided mode is possible, which is done numerically. The highest n_{eff} that fulfills (A.18) gives the fundamental mode of the waveguide structure.

Eigenmodes can be calculated by inserting the so found n_{eff} into the matrix Δ and propagating the 4-vector through the structure using Eq. (A.15). The correct starting condition, the coefficients a and b, can be obtained from Eq. (A.17), and z-components are given by Eqs. (A.11) and (A.12).

Curriculum Vitae

Full name:	Wolfgang Georg Scheibenzuber
Birth:	October 3rd 1984 in Vilsbiburg, Bavaria, Germany
Secondary education:	Maximillian-von-Montgelas Gymnasium Vilsbiburg Abitur (1.0)
Scholarships:	Bayerische Begabtenförderung (BayBFG) Elite-Netzwerk Bayern Studienstiftung des Deutschen Volkes

04/2005–08/2005	Internship at BMW Regensburg, damage analysis
10/2005–08/2009	Studies at Regensburg University, diploma in Physics (1.0, with honors), diploma thesis: "Optical Gain in (Al,In)GaN Laser Diodes"
10/2009–09/2011	Doctoral studies at Freiburg University, scientific assistant at the Fraunhofer Institute for Applied Solid State Physics (IAF), doctoral thesis: "GaN-Based Laser Diodes: Towards Longer Wavelengths and Short Pulses"

W. G. Scheibenzuber, *GaN-Based Laser Diodes*, Springer Theses,
DOI: 10.1007/978-3-642-24538-1, © Springer-Verlag Berlin Heidelberg 2012

Journal Papers

W.G. Scheibenzuber, U.T. Schwarz, R.G. Veprek, B. Witzigmann, A. Hangleiter, Calculation of optical eigenmodes and gain in semipolar and nonpolar InGaN/GaN laser diodes. Phys. Rev. B **80**, 115320 (2009)

W.G. Scheibenzuber, U.T. Schwarz, T. Lermer, S. Lutgen, U. Strauss, Antiguiding factor of GaN-based laser diodes from UV to green. Appl. Phys. Lett. **97**, 021102 (2010)

W.G. Scheibenzuber, U.T. Schwarz, L. Sulmoni, J.-F. Carlin, A. Castiglia, N. Grandjean, Bias-dependent absorption coefficient of the absorber section in GaN-based multisection laser diodes. Appl. Phys. Lett. **97**, 181103 (2010)

W.G. Scheibenzuber, U.T. Schwarz, R.G. Veprek, B. Witzigmann, A. Hangleiter, Optical anisotropy in semipolar (Al,In)GaN laser waveguides. Physica Status Solidi. C **7**, 1925 (2010)

W.G. Scheibenzuber, C. Hornuss, U.T. Schwarz, Dynamics of GaN-based laser diodes from violet to green. Proc. SPIE **7953**, 79530K (2011)

W.G. Scheibenzuber, U.T. Schwarz, Polarization switching of the optical gain in semipolar InGaN quantum wells. Physica Status Solidi. B **248**, 647 (2011)

W.G. Scheibenzuber, U.T. Schwarz, L. Sulmoni, J. Dorsaz, J.-F. Carlin, N. Grandjean, Recombination coefficients of GaN-based laser diodes. J. Appl. Phys. **109**, 093106 (2011)

W.G. Scheibenzuber, U.T. Schwarz, Fast self-heating in GaN-based laser diodes. Appl. Phys. Lett. **98**, 181110 (2011)

W.G. Scheibenzuber, C. Hornuss, U.T. Schwarz, L. Sulmoni, J. Dorsaz, J.-F. Carlin, N. Grandjean, Self-Pulsation at zero absorber bias in GaN-based multisection laser diodes. Appl. Phys. Exp. **4**, 062702 (2011)

W.G. Scheibenzuber, U.T. Schwarz, T. Lermer, S. Lutgen, U. Strauss, Thermal resistance, gain and antiguiding factor of GaN-based cyan laser diodes. Physica Status Solidi. A **208**, 1600 (2011)

W.G. Scheibenzuber, U.T. Schwarz, L. Sulmoni, J.-F. Carlin, A. Castiglia, N. Grandjean, Measurement of the tuneable absorption in GaN-based multi-section laser diodes. Physica Status Solidi. C **8**, 2345 (2011)

P. Perlin, K. Holc, M. Sarzynski, W.G. Scheibenzuber, L. Marona, R. Czernecki, M. Leszczynski, M. Bockowski, I. Grzegory, S. Porowski, G. Cywinski, P. Firek, J. Szmidt, U.T. Schwarz, T. Suski, Application of a composite plasmonic substrate for the suppression of an electromagnetic mode leakage in InGaN laser diodes. Appl. Phys. Lett. **95**, 261108 (2009)

J. Rass, T. Wernicke, W.G. Scheibenzuber, U.T. Schwarz, J. Kupec, B. Witzigmann, P. Vogt, S. Einfeldt, M. Weyers, M. Kneissl, Polarization of eigenmodes in laser diode waveguides on semipolar and nonpolar GaN. Physica Status Solidi. RRL **4**, 1 (2010)

T. Lermer, M. Schillgalies, A. Breidenassel, D. Queren, C. Eichler, A. Avramescu, J. Müller, W.G. Scheibenzuber, U.T. Schwarz, S. Lutgen, U. Strauss, Waveguide design of green InGaN laser diodes. Physica Status Solidi A **207**, 1328 (2010)

T. Lermer, A. Gomez-Iglesias, M. Sabathil, J. Müller, S. Lutgen, U. Strauss, B. Pasenow, J. Hader, J.V. Moloney, S.W. Koch, W.G. Scheibenzuber, U. T. Schwarz, Gain of blue and cyan InGaN laser diodes. Appl. Phys. Lett. **98**, 021115 (2011)

S. Lutgen, D. Dini, I. Pietzonka, S. Tautz, A. Breidenassel, A. Lell, A. Avramescu, C. Eichler, T. Lermer, J. Müller, G. Brüderl, A. Gomez, U. Strauss, W.G. Scheibenzuber, U.T. Schwarz, B. Pasenow, S. Koch, Recent results of blue and green InGaN laser diodes for laser projection. Proc. SPIE **7953**, 79530G (2011)

J. Dorsaz, D.L. Boiko, L. Sulmoni, J.-F. Carlin, W.G. Scheibenzuber, U.T. Schwarz, N. Grandjean, Optical bistability in InGaN-based multisection laser diodes. Appl. Phys. Lett. **98**, 191115 (2011)

Conference Contributions

W.G. Scheibenzuber, U.T. Schwarz, R.G. Veprek, B. Witzigmann, A. Hangleiter, *Calculation of Optical Eigenmodes and Gain in Semipolar and Nonpolar InGaN/ GaN Laser Diodes* (ICNS-8 in Jeju, South Korea, 2009)

W.G. Scheibenzuber, U.T. Schwarz, *Invited Talk: Polarization Switching of the Optical Gain in Semipolar InGaN Quantum Wells* (E-MRS fall meeting in Warsaw, Poland, 2010)

W.G. Scheibenzuber, U.T. Schwarz, T. Lermer, S. Lutgen, U. Strauss, *Antiguiding Factor of GaN-based Laser Diodes from UV to Green* (IWN 2010 in Tampa, Florida, USA, 2010)

W.G. Scheibenzuber, U.T. Schwarz, L. Sulmoni, J.-F. Carlin, A. Castiglia, N. Grandjean, *Bias-dependent Absorption Coefficient of the Absorber Section in GaN-based Multi-Section Laser Diodes* (IWN 2010 in Tampa, Florida, USA, 2010)

W.G. Scheibenzuber, U.T. Schwarz, T. Lermer, S. Lutgen, U. Strauss, *Dynamics of GaN-based Laser Diodes from Ultraviolet to Green* (SPIE Photonics West 2011 in San Francisco, California, USA, 2011)

W.G. Scheibenzuber, U.T. Schwarz, L. Sulmoni, J. Dorsaz, J.-F. Carlin, N. Grandjean, *Auger Recombination in GaN-based Laser Diodes* (ICNS-9 in Glasgow, Scotland, 2011)